101 QUESTIONS
YOUR CAT WOULD ASK ITS VET

101 QUESTIONS
YOUR CAT WOULD ASK
ITS VET
IF YOUR CAT COULD TALK

Bruce Fogle

Illustrated by Lalla Ward

Carroll & Graf Publishers, Inc.
New York

Copyright © 1993 by Dr. Bruce Fogle
Illustrations copyright © 1993 by Lalla Ward

Published by arrangement with Michael Joseph, a division of the Penguin Group, London.

First Carroll & Graf edition 1993

Carroll & Graf Publishers, Inc.
260 Fifth Avenue
New York, NY 10001

Library of Congress Cataloging-in-Publication Data

Fogle, Bruce.
    101 questions your cat would ask its vet if your cat could talk /
  Bruce Fogle : illustrated by Lalla Ward.—1st Carroll & Graf ed.
        p.   cm.
    ISBN 0-88184-952-9 : $7.95
    1. Cats—Miscellanea. 2. Cats—Health—Miscellanea. 3. Cats-
  -Behavior—Miscellanea. I. Title. II. Title: One hundred one
  questions your cat would ask its vet if your cat could talk.
  III. Title: One hundred and one questions your cat would ask its vet
  if your cat could talk.
  SF447.F64 1993
  636.8—dc20   93-9185
                                                            CIP

Manufactured in the United States of America

Contents

Acknowledgements

In a survey of what cat owners want most from their veterinarians, the unexpected answer was 'someone who listens'. Pet owners were more than satisfied with their vet's technical abilities but less so that they were being listened to and being given the kind of advice they most wanted.

The first veterinarian I worked for, Brian Singleton, taught me the importance of listening to clients. What I learned was that although people are seriously concerned for their cat's well-being there is almost always a sense of fun in the relationship. The questions that follow are among the most common I am asked. My thanks go to Brian and to the thousands of cat owners I have met during the last two decades, who together have made 'work' such a pleasure. Thanks, too, to the cats.

Introduction

I admit it. The questions that follow are transparently anthropomorphic. Cats aren't interested in explanations. They don't want to know whether neutering will affect their personality, if acupuncture will alleviate pain, why their teeth chatter when a bird lands on the windowsill or why they spray the walls with urine when the new baby is brought into their home. They're simply interested in comfort, security, mental and physical stimulation and, most of all, survival. Cat people, on the other hand, are always asking vets questions. So why should I put words in cats' mouths?

The simple answer is because it is fun, but there is a subtler reason, too; my family's relationship with our cat, Millie, exemplifies it. My wife, Julia, had never lived with a cat. She was raised with dogs, ardently loved dogs and may well have married me so that she would have in-resident veterinary assistance for her own dog.

After Millie, a silver tabby kitten, arrived in our home, I would come back from work to find Julia nestled in a corner of the sofa with a sleeping kitten on her lap. When asked what she did that day, Julia would reply that she hadn't managed to get out because Millie:

(a) needed someone to play with;
(b) wanted to go outside and needed to be watched;
(c) was sleepy and needed a lap to curl up on.

Within a few weeks, a strong emotional bond had developed between them. Millie looked upon all of us, and Julia in particular, as providers of food, affection, warmth and security. In return, we unconsciously made an emotional investment in Millie. She became distinct, unique; at some undefined moment she evolved from a 'cat' into a 'personality'. We started to understand her emotions and read her feelings. We became better at communicating with her. We developed a rapport, a mute form of communication.

Unless we try to be coolly and dispassionately academic, we develop a care and dependency relationship with cats; as a species, we have a life-long need to nurture, so it is only natural that such a relationship develops. I know that domestic cats are a species of highly evolved carnivores whose needs are different from humans', and that if they were large enough, they would look upon us as meals rather than parents, but on a personal and purely emotional level I see my cat as a unique individual for whom my family is responsible.

When we see our cats in this way, it means that we develop an understanding and respect for their feelings and emotions. Having the cats ask the questions in this book acknowledges the fact that most of us treat them as individuals, as members of the family with their own unique and personal needs.

The cat happens to be an independent species that has chosen to live with us. By moving into our homes and gardens, cats provide us with a window on nature, front row seats into an exciting other world. I hope that by having cats ask these questions, by being intentionally anthropomorphic, I can reinforce the fact that all intelligent animals – cats, dogs, foxes, raccoons, deer – share a variety of needs and emotions, feelings and desires that we understand because we have them too.

Some of the questions are simply tongue-in-cheek. If you keep score as you read the questions, you will be able to assess how 'human' you think cats are. A score of no 'tongue-in-cheek' questions means you think cats are small soft fur-people. A score of 101 tongue-in-cheek questions means that you think cats are machines that live their lives on auto-pilot. You will notice that the answers are orientated towards the owners rather than the cats. That is because very few cats I have met can read. But you can recite to them. They are magnificent and knowledgeable listeners.

Finally, you might think that your cat would like to ask a question. Whatever it may be, please write it down and send it to me:

Bruce Fogle
Cat Questions
c/o Michael Joseph Ltd
27 Wrights Lane
London W8 5TZ.

I'll try to include it in future editions.

CHAPTER ONE

Instincts and Communication

1. I like to think I have a good range of voices. I can hiss, spit, purr, growl and chirp, but why is it that sometimes when I meow people don't seem to hear me?

Considering their size, cats have a magnificent range of noises, certainly wider ranging than those of their confrères, dogs. The hiss is almost snake-like: when a tabby cat is curled up and hisses, she might even be mistaken by a predator for a snake. The spit accomplishes two functions: the noise itself can be frightening; at the same time, the animal that has been spat on smells hot cat breath, on its own enough to frighten many predators away. Purring is a tension-reducing sound. Cats purr when they are relaxed or when they *want* to relax. Rumbling growls are audible sounds of anger; chirps are sounds of delight.

Meows, on the other hand, have specific meanings according to their duration, pitch and inflection. Cats understand different meows instantly, but people need training before they can interpret whether the meow signifies impatience, worry, consternation, gentle request or demand. People have an added problem, however. Cats have a hearing range, and a meow range, that is beyond the hearing range of humans. When a cat utters a 'silent meow' people don't respond because they just don't hear it. If they see the cat meow, they are enchanted and think the silent meow has been performed especially for them. They don't understand that the meow is silent only because their hearing is so limited. It is a perfectly good meow to any other cat.

1

2. I normally say 'hello' by raising my tail and sticking my behind in someone's face. Why do people find this unpleasant?

Odour plays a major role in communication between cats. But just as important as the odour is the willingness of one cat to allow another to capture his scent. This is one of the reasons why, when cats meet for the first time, they circle each other. What they are trying to do is to capture the scent produced from two different glands around the other's behind. Cats that have just met haven't had time to work out seniority, so they keep their tails down, protecting their perianal glands and anal sacs from inspection, and move in a circular pattern as each tries to sniff the other. The dominant cat eventually gets first sniff.

When cats meet other cats or people they know, there is no need to hide their scent, so they carry their tails high and present their posteriors for inspection. In this way, they are simply acknowledging that the other cat or person is dominant or 'top cat'.

In cat hierarchy, people replace parents or, to be more specific, people replace mother cats as providers of food, comfort and security. Mother cats routinely lick clean their kittens' bottoms, so it is natural to the cat to present her backside to her people parents. However, because bums in faces have a different connotation in human behaviour, people find this form of greeting unpleasant.

3. Why do I feel more content and relaxed with strange people than I ever do with strange cats?

In natural circumstances, where they have to survive on their own, cats live isolated lives. They hunt on their own, sleep on their own and amuse themselves without the need for companionship from other cats.

There are only two occasions when cats come together. The first is when there is a large source of food constantly available. In these circumstances cats will live communally, simply because there is no need for intense competition for food. However, there is still competition for territory and for the second reason cats come together, sex. When the food source is not man-made but consists of rodents and birds, the only time during the year that one cat comes in physical contact with another is when they have sex.

In most circumstances, another cat means competition. A stranger is an intruder who might take over the territory, and the food and the sexual rights that go with it. People, on the other hand, provide cats with food and warm, comfortable homes, thereby establishing social bonds with cats, but they are different enough so that they are not in competition with cats for the same food source, for sex or for territory. They seldom, for example, feel the need to mark out their territories in the way cats do. The result is that cats don't feel as intimidated by strange people as they do by strange cats.

4. I'm quite reserved and find it difficult to show my emotions. How can I tell people I love them?

Cats don't wear their emotions on their sleeves the way dogs and people do. Both of those species have dramatic and blatant ways of showing love and affection. People smile, lower their heads and voices and mutter love mutter. Dogs smile, lower their heads, wag their tails and mutter love mutter.

Cats are more subtle. Because they are not pack animals, there was never the need to develop a sophisticated range of 'come hither' signals. Cats are brilliant at saying 'I hate you', but less obvious at proclaiming their love. Some will do so vocally, with a quick and sometimes repetitive 'chirp'. Others will purr, although people often mistake purring to reduce anxiety with the purring of contentment.

Face rubbing is another gentle love touch. Although it is an odour-marking technique, cats show affection for certain people by rubbing their faces against human faces. They also body-rub and tail-wrap.

Curling up on a warm lap is one of the grandest ways to show affection. Even cynical people, who think that cats do this only because it is the warmest place to sleep, gradually melt when they see such feline contentment.

Finally, gift giving shows that cats care. People can find it disconcerting when a cat brings in a captured trophy, but it is a generous, even loving, gesture.

5. When I'm being petted, I lapse into a reverie of delight, but sometimes I get this overwhelming urge to bite my human. In fact, sometimes I do. Do I have a personality disorder?

Petting creates an emotional conflict for some cats. As kittens they rely upon their mothers for warmth, nourishment and hygiene. While the kitten snuggles in to suckle, his mother cleans through his fur with her rasp-like tongue. She licks his ears and bottom and grooms his fur, removing parasites and any debris that has accumulated. It is one of the first sensations of life and one of the most enjoyable for the kitten.

Social grooming continues as long as the kitten suckles. In natural cat colonies, where mother and daughters remain together in a lion pride type of situation, the mother might continue to groom her now grown kittens occasionally, but she does so for relatively short periods of time.

Once a kitten leaves home, however, physical contact with other cats usually ends. Cats don't normally groom each other. They maintain physical contact only to keep warm, to have sex or when ongoing and social bonds allow them to do so.

There is only one significant exception to this. In captive situations kittens are sometimes kept together in human households. They are neutered at an early age, usually before they reach puberty. Under these circumstances, physical contact between cats continues. They sleep together and continue to groom each other.

5

The 'culture' of these cats is quite different to the 'culture' of their brethren who lead a feral existence. In the wild, social grooming only occurs between related females.

People, on the other hand, continue social grooming throughout their lives. Touch is their most important sense, and they get pleasure out of petting soft, warm and sensuous cat fur. Cats like it too, initially, because it is reminiscent of their mother's grooming. But when it continues for too long, it creates an emotional conflict. The cat feels a sudden threat, bites, jumps down, feels better and, more often than not, hops back onto his human's lap and allows himself to be petted again. The problem is not a feline split personality, but rather the human's need for constant contact comfort.

6. Night-time excites me. Why is it that at dusk I feel this urge to march through my cat flap and go native?

When allowed to live their own lifestyles, cats settle in to the routines for which they are best equipped. Their senses of smell, hearing and sight benefit from the conditions at night. The scent of rodents, the mainstay of the natural menu, is greater when the ground is slightly damp, which it is likely to be in the dew of the night. The cacophony of daytime sounds diminishes as dusk arrives, allowing cats to hear rodents' high-pitched squeaking sounds, which are inaudible to the human ear. Also, cats can see in only one-sixth the amount of light needed by humans.

Cats rest and conserve energy during the daylight hours, while many of their natural enemies, such as dogs and humans, prefer to be active when it is light. Although cats are efficient hunter-killers, they are small and so are constantly at risk from larger carnivores. The risk is lessened slightly if they don't hunt at the same time as their enemies.

Because they have only just moved from the African savannah into people's homes, from an evolutionary viewpoint, regardless of how domesticated they might appear to be, even the most elegant and fluffy cat still feels the call of the wild and the urge to go hunting when light fades at the end of the day.

Finally, just like people, as night falls cats, too, have an urge to go cruising for sex.

7

7. I am a lethal bird killer. Will I ever run out of victims?

Not likely. Although domestic cats have been in Europe for a few thousand years and have been in the Americas for only several hundred, there have always been other land-based bird predators in these regions.

Most cats make poor bird hunters. Farm cats far prefer to concentrate on rodents, which are plentiful and quite tasty. But even docile house cats have the potential to become superb killing machines and some become specialist bird cats. In many areas cats are the main predators of birds, although this does not necessarily adversely affect bird populations.

In legislatures throughout the world, birds have their advocates. In Australia, for example, there is at least one jurisdiction in Victoria in which cats must return home at sunset to put on leads and collars, which they must wear until dawn. Legislators feel that cats on leads at night will not kill birds. In Norway cats must measure the distance they travel. They may go birding only within 100 metres of home, any further away and they become legal prey for humans.

Cats and birds are natural predators and prey respectively. Supreme hunters though cats may be, over the centuries they have not made a significant dent in the population of their favoured prey, rodents. Because they kill selectively, choosing young, elderly, debilitated or unwary birds, it is unlikely that cats will ever compete seriously with humans in their ability to drive animals to extinction.

8

8. I believe in sharing. Why do people get so upset when I jump on the bed in the morning and give them a fresh corpse?

Dawn is one of the best times for successful hunting. Dew worms are still lying on the ground. Frogs are concentrating on flies. Flies are concentrating on dog droppings. Birds are concentrating on dew worms. Rodents are returning to their nests. There is a cornucopia of food, a three-star restaurant fresh for plundering. The only complication is that house cats usually have full stomachs. They have no physical need to hunt but they simply can't help it. They follow their instincts. The question then is what to do with the prize.

Many cats remain in arrested emotional development when it comes to capturing, killing and eating prey. They remain kitten-like in that they capture, release and capture their prey again and again. They torment and tease rather than carry out the natural and instantaneous death bite that feral cats would inflict on prey almost immediately after catching it. Domestic cats with full stomachs catch animals because of the thrill of the hunt, much in the same way that fishermen catch and release fish.

In a curious way, adult household cats in particular have split personalities. Part of their development is arrested because people feed them and house them, so they retain some kitten-like behaviours. This is part of the reason why they torture prey rather than inflicting instantaneous death bites.

Adult behaviours mix with these retained juvenile

9

actions and account for the cat's willingness to be thoughtful and giving.

Because domestic cats develop social bonds with people, they often want to share their prizes with them. Dawn is the most successful hunting time; it is also the time of day when people are most likely to be in bed. A wounded frog or bisected earthworm on the pillow at dawn is simply a manifestation of feline generosity; it is only because people do not show generosity in a similar way that they find it unpleasant.

9. I have this compulsion to stalk birds, but whenever I do the tip of my tail twitches and they fly away. How can I improve my technique?

Domestic cats hunt with patience and stealth. They often go back to locations where previously they have been successful. There they hide in long grass, waiting for a rodent to come out of its hole in the ground. The experienced hunter waits silently and perfectly still, allowing the rodent to move away from the hole before pouncing on it.

When stalking birds, cats use the same slow, steady technique. They hide in deep grass, and move forward either in slow motion or in short bursts at a time. Because most birds have widely placed eyes, they have superb peripheral vision. Many birds have 360-degree vision, but the stealthy cat uses long grass as his hide and keeps perfectly still. This allows the patient stealthy cat to creep near enough to the bird to pounce on it.

Domestic cats face a serious problem, however, for rather than having savannah grass to hide in, they more often have a perfectly mown lawn. Even worse, their humans might have kitted them out with bells on their collars. Many cats easily learn how to keep even multiple bells silent as they stalk forward, but they still have no place to hide. This creates frustration which they show by twitching the tip of their tail. The only guaranteed way to improve technique is to convince people to stop cutting the grass.

10. If I must have a feline companion, which sex is best?

This answer varies with the circumstances in which cats find themselves. Let's start with free-living town or farm cats. If cats eat, sleep and breed on their own, independent of people, they naturally divide up according to sex. Females club together, forming a segregated colony. They will care for each other's young, even suckling them. They will allow their daughters to remain within the colony, but will drive away their sons when they reach puberty. Most of these males will create an informal alliance with each other, a feline brotherhood. Their degree of co-operation is at its most dramatic when they form a quiet line, waiting to mate with a female in season.

This natural behaviour doesn't translate easily into the circumstances in which most cats find themselves. When cats live with humans, two of the most important variables are compromised. The first is territory. Many cats find themselves restricted to territories demarcated by humans rather than those they have staked out themselves. Human territories are almost always considerably smaller than cat territories, often consisting of no more than a postage stamp-sized back garden. In fact, many human territories have no outdoor space at all. The second variable is sex. For various reasons, most domestic cats end up desexed. This has a considerable effect upon whom a cat wants to live with.

Sexually mature entire (unneutered) cats can look upon the intrusion of another cat of the same sex into their artificially small human territory as a great challenge. This is why it is always best for cats to meet each other before they have reached puberty, when it is more likely that they will eventually become companions regardless of their sex. If cats have matured, it is better to be introduced to a member of the opposite sex. If, as is more likely to be the case, a cat has been neutered, then the sex of the newcomer is less important. Neutering is a great leveller. Although male and female cats have considerably different behaviour patterns, when they are neutered their behaviour is much more similar, and most like that of the unneutered female. If in doubt it is best to live with a member of the opposite sex.

11. People worry that I'm going to smother their new baby. How can I convince them that human babies don't interest me?

This is one of the greatest concerns that people have about cats. They worry that their cat will be jealous of the new baby and will try to scratch him or, more benignly, that she will find he is a convenient sleeping companion.

Cats like routine and can be upset when their routine is altered. A cat can show her lack of interest in the baby simply by hiding from him, but people should be aware that showing curiosity doesn't mean that a cat is planning something evil.

Ideally people should allow their cat to sniff the new scents that the baby has brought home. There is never any need for a cat to sleep in or even to enter the nursery. If, previously, this room was her sleeping quarters, it should be closed off well before the baby arrives. The lack of signs of feline aggression should be enough to reassure the worried human.

People can make sure that the cat won't sleep with their baby by hanging netting around the crib. In return, by sticking to normal sleeping patterns, the cat can reassure her humans that she has no felonious intent.

12. I'm told my teeth sometimes chatter, my feet paddle and my eyelids flutter when I sleep. What's happening?

That is cat dreaming. And cats do it a lot.

Of all the animals that live with people, cats sleep most. Given the slightest opportunity a cat will take a cat-nap. Most cats nap so much they sleep twice as much as people do. But most of this sleeping time is spent in light sleep in which they don't dream.

Only when a cat is completely relaxed will he fall into deep sleep. Electrical activity in the brain changes and the excitement begins. Under closed eyelids the eyes move rapidly. Whiskers can twitch, jaws move and claws and paws extend and contract. When people see this happening they usually smile and call it 'chasing rabbits'.

These electrical and physical changes occur when people dream, but the question remains why cats dream at all, and if they do, what they dream about. Almost all mammals dream while they are asleep. Only marine mammals like dolphins and whales, and the occasional land mammal such as the Australian spiny ant-eater, don't have the electrical brain changes associated with dreaming. Cats are very intelligent animals, and their brains are constantly taking in tremendous amounts of information. One of the possible reasons for dreaming is that it is the cat's way of reclassifying or even getting rid of the build-up of unnecessary data in the brain.

15

13. If I climb a tree can I be sure I'll ever be able to get back down?

Just as cats are unique in that they sleep more than any other domestic animal, they are also unique in their ability to live in a vertical as well as a horizontal world. Mature cats can jump up to seven times their own height. All eighteen claws are equipped with special strong muscles that extend the claws and allow them to act as grappling irons. Finally, extra fast transmission of information from the organs of balance in the ears into the brain means that cats can virtually walk tightropes.

These superb and certainly superhuman abilities mean that cats are extremely accurate when they jump and rarely miss their footing. A healthy cat will take a flying jump at a tree, land exactly where she has planned to land, and then effortlessly grabble her way up to the branch she wants to reach. From this vantage point she can survey her territory for food or intruders, or wait until an unsuspecting bird arrives in a nearby branch.

A cat in a tree is a contented, happy cat, because in the cat hierarchy, height means seniority. The cat that captures the high ground is the dominant cat in any demarcation disputes with other cats. Of course, she is also safer from her natural predators. The question remains how she will get down.

First-time climbers can find this difficult. Grappling claws don't work well backwards. It takes some trial and error before a cat develops the tail-first descent technique. In the meantime, people shouldn't panic. There has never been an authenticated report of anyone finding a cat skeleton in a tree.

16

14. How good is my memory?

Some people think that cat behaviour is all a matter of instinct. Cats behave the way they do only because of the genes they have inherited. Not true.

Cats are, of course, restricted by their genetic inheritance just as humans are. Cats will never learn algebra, and people will never hear high frequency mouse squeaks. But that doesn't mean that a cat's memory is any less sophisticated than a human's.

Take the cat basket, for example. Many people know that at the mere sight of the cat basket their cat suddenly disappears from the face of the earth. The cat remembers that seven years ago, the last time the cat basket was used, he was taken to the vet's and he hated the experience. Better still, have two people sit down and simply discuss taking the cat to the vet. The cat's memory can be brilliant here, too, for merely hearing certain buzz words can remind him of past unpleasant experiences.

People make their greatest mistake in failing to realise what magnificent observers cats are. Cats monitor their human's body language and remember it. They soon learn their human's routine and expect that routine to be maintained. But when the routine alters, they recognise that, for better or worse, their lifestyle is going to be altered too. It might have been years since the previous event, but cats remember it as if it were yesterday.

17

15. I enjoy television, especially sports programmes, but why is it so difficult for me to catch the ball?

Cats have an advantage over dogs. All over the world, in any format, they can watch television. Throughout Europe, Australasia and the Americas, bored indoor cats amuse themselves by watching golf, snooker and tennis. Some will sit in front of the TV and, like a metronome, will look left then right as the movement goes from one side of the screen to the other.

Bored American dogs, on the other hand, can't pass the time in a similar pursuit. The American format of television transmission is slower than elsewhere in the world, and the result is that American dogs only see hundreds of dots on their screens. They can listen to sporting events, but if anything exciting happens they almost invariably go behind the set to see where the noise is coming from.

Cats aren't too interested in whether people have colour or black and white television because colour is unimportant in life. They are more interested in what type of aerial is used. They prefer satellite and cable for clarity. However, only cats who know how to use the stop-action button on the video remote control ever catch the ball.

16. *Why do I race down the stairs in front of my humans, then collapse in a heap so they have to step over me?*

Humans can be quite boring playmates. They often don't realise that cats, especially young cats or members of brazen breeds like Siamese, Burmese and Abyssinians, frequently need to flex their bodies and their minds. Cats try to entice humans into their games, but humans often have other things on their minds and don't respond.

A general rule is that humans are more responsive when they are sitting or lying down than when they are standing or moving. The 'collapsing in a heap' ploy seldom works because long-legged humans find it easy to step over the feline obstacle.

There are other general rules, too. The larger the family is, the less time is spent playing with the cat. If a cat wants the attention of adult humans, he should make certain there are no children around. If, however, he lives with a single human, he will, curiously, get more attention from someone who goes out to work each day than from someone who doesn't work or who works from home. Out of guilt, love or boredom, working humans are more likely to respond to the 'cat in a heap' ploy than any others.

17. Why do I sometimes lash my tail, lower my head and growl at nothing?

It might be nothing to the human observer, but it is definitely something to the cat. First of all, it can be pain. If a cat feels sudden pain, whatever the source, instinctively she will act angry and growl. Even healthy young cats occasionally feel a twinge of pain in the intestines or in the urinary system, from tooth decay or from overfilled anal glands.

Some cats, however, hiss and growl at what seems to be nothing, and this raises the question of whether they hallucinate. Certain plants, such as catnip and spider plants, contain chemicals that can cause feline hallucinations, but even without partaking of these some cats act as if they see things, and don't like them. It may well be that cats are uniquely different to other species, but as it is usually housebound cats that behave this way, it is more likely that the behaviour is a manifestation of boredom than of anything else.

Veterinarians should always examine cats who see things, to eliminate any medical reasons for the behaviour. If no physical reasons are found, hallucinators should be given more mental stimulation as a method of draining off mental activity that cannot find a natural outlet.

CHAPTER TWO

Emotions and Behaviour

18. I am a Siamese. Each time I catch sight of the new cat my people have brought into the house I want to claw his eyes out. Why do I feel such rage?

Personality varies from cat to cat, but certain personality traits are more common in some breeds than in others. Siamese are more demonstrative with their emotions than most other breeds, and they are the breed least likely to accept feline strangers into their home.

Cats willingly live with each other under certain conditions. Most important of all is early contact with the potential cohabitant. If cats are raised with one another from kittenhood, they will not only live together peacefully, but they will also, probably, actively enjoy each other's company. Siamese cats in particular will work as a team, often sleeping together, playing together and even, on occasion, hunting together.

This is not the case when they meet as adults. By then a cat has an established routine. She has trained her humans to perfection, and feels jealous when another cat enters her space and her life. The resident cat might try to drive the other out of her territory. The problem is that humans have locked the doors and the newcomer can't escape.

To avoid this happening people should introduce a new cat into the home gently. He should be restricted to one room while the resident, especially if she is a Siamese, has free movement everywhere else. The resident should be allowed to investigate the newcomer, not the other way around. Doing this reduces, but does not eliminate, the resident's desire to perform eye surgery on the interloper.

19. Although I usually keep my cool, why do I suddenly feel compelled to race around the room and do 'the wall of death'?

Compared to humans, and especially compared to dogs, cats are regal and dignified. They don't grovel. They don't pander. They don't show unnecessary emotion. Cats are always in control, when confronted with danger, in their relationships with others, when hunting and when relaxing.

It is difficult to keep that control because it is all a sham. While on the outside a cat is showing steady calm, inside is another matter. When confronted with danger, adrenalin pumps around the body, the heart rate increases, the blood pressure changes and the skin temperature goes up. But all the while the cat doesn't give any indication of his true feelings.

When the danger passes, the body returns to normal, but there has been no chance to vent all of that emotion – even the most controlled cat needs to let off steam. Hunters will do it after a capture and kill. They dance around the body in a reverie of excitement and joy. Housebound cats never have the chance. Instead, house cats have a 'mad half hour'. They release their pent-up energy by suddenly racing around the house. Many enjoy using furniture. They charge across the carpet, then up and over a chair or sofa, and off into another room. Others race in circles in a room until they reach critical velocity. At that point they defy gravity and, using centrifugal force, race around the walls as far as possible. Dignity returns a few minutes later.

20. I think I'm a pretty good mother, but lately my kittens have been getting on my nerves. I find myself constantly shouting at them and disciplining them. What's happening to me?

Just as people go through a stage where their children get on their nerves, mother cats do too. For people, this stage usually occurs when the children are between twelve and eighteen years old, but with cats it happens when the kittens are twelve to eighteen weeks old.

Maternal care is of paramount importance if kittens are going to survive. Tom cats rarely show any interest in the litters they have sired, but mothers must feed, comfort and protect. In doing so they form warm caring relationships with their kittens. A mother, frightened for the safety of her litter, is the most dangerous of all cats.

If the social bonds between a mother and her kittens were to continue, she would never want them to leave and become independent. Equally important, they would never want to leave. So, as her milk dries up and her female hormone levels alter, her need to care for her progeny diminishes. At the same time, their play becomes increasingly aggressive. Their needle-sharp teeth and claws hurt her, and she has every justification to rant and rave. Her male kittens are the most aggressive in their play combat with her, and she responds in kind. Soon, realising that mum is no fun any longer, they pack up and leave. In the wild when food is scarce her daughters eventually do the same, leaving her with her own hunting territory, and the freedom to devote all her attention to her next litter.

21. After my mother died I became withdrawn and nervous, now everything frightens me. Will I get over it?

There is little doubt that cats grieve. And because people frequently act as cat substitutes for other cats, grief can be felt for the loss of human companionship too.

Cat grieving has been investigated by veterinarians and even by human psychiatrists. It is more likely to occur in dependent cats, those that enjoy being cared for and looked after. Typically, the bereaved animal seems disorientated. He searches for his missing companion, and appears restless and more vocal.

Grieving cats often stop eating. They play less and might stop grooming themselves. If they are elderly, this can be misinterpreted as old age kidney failure or even senile dementia. This is a dangerous misinterpretation, for grieving cats might simply be pining, and if that is the case, the problem can be overcome.

Cats can get over their grief if people accept that special treatment is necessary. Once a veterinarian has eliminated any medical reasons for the cat's apparent depression, the cat should be handled frequently, spoken to quietly and groomed regularly. Tasty nutritious foods, especially foods with strong odours, should be offered and hand-fed if necessary. New routines must be created in which the bereaved cat develops an emotional relationship with his replacement companion.

22. Will having a kitten for company help me in my bereavement?

Not straight away. If a cat is grieving the loss of companionship, a new kitten simply isn't a suitable replacement for that loss. The loss has been unique. Nothing can replace it. What can happen is that with time the grieving cat can form a relationship with another cat but that will happen only after new and different bonds develop.

Loneliness is not as great a problem for cats as it is for people or for dogs. In fact, in many two-cat households, when a cat dies, the survivor might virtually jump for joy. The loss is actually a gain in which the survivor now has the territory to himself. People often notice that the survivor develops a more outgoing temperament, becomes more demonstrative, affectionate and vocal. Weight is gained, the coat improves and confidence increases.

If this is the case, there is little or no value to the cat in having a new kitten introduced into his home. If, however, his humans want another cat, then it is best for them to get a kitten rather than another adult cat, and to introduce it along the lines that I have described in question 18.

23. Lately I've felt washed out, listless and dull. I've had no energy. All I want to do is sleep. Is this what people call depression?

Probably not. Cats do suffer from depression, and then they become vacant and uninterested in life. They may stop grooming themselves and refuse food. Any serious change in routine can bring on possible cat depression, but when a cat's personality changes, people should be concerned first with the probability of physical illness.

These are serious changes, and cats that suffer from them should be examined by a veterinarian. The problem might be as simple as a fever caused by a circulating infection picked up in a cat fight. Antibiotics will bring about an almost instantaneous recovery within a day.

A whole range of other more serious medical conditions, such as kidney or liver disease, slow-acting virus infections, anaemia and a variety of cancers, all produce similar clinical signs of lassitude. Each requires an accurate diagnosis before treatment is undertaken.

When other medical reasons for the behaviour change have been eliminated, then depression can be assumed to be the cause. This is treated in a wholly different way. Although anti-depressant drugs can sometimes be used successfully, gentle nursing is the preferred treatment. Depressed cats need attention. If they have stopped grooming themselves, humans should take over and keep their fur in good condition. If they have stopped eating, they should be hand-fed nutritious and pleasant-smelling food. With tender care, most cats will come out of depression without the need for drugs.

24. I have an insatiable need to eat houseplants. My human recently bought a spider plant and I denuded it in a week. Do I have what some cats call an addictive personality?

There might be a genuine addiction involved, but it would be a typical behaviour rather than a sign of a true personality disorder.

Although they are true carnivores, most cats will eat grass naturally. They eat it to help them to vomit, but also because they sometimes simply enjoy having a salad. Those that lead an outdoor life will develop preferences for certain grasses and will often favour succulent young grass to drier older forms. Indoor cats don't have the advantage of choice, and this is when they turn to houseplants.

Eating houseplants can be dangerous. Some cats, perhaps most, can't differentiate between safe and harmful ones. They chew greens such as ivy leaves, and vomit afterwards because they have mildly poisoned themselves. Worse, they might chew on dangerous plants, such as Dieffenbachia, and suffer severe poisoning. Spider plants, however, are safe and intriguing.

House cats are attracted to spider plants in much the same way that outdoor cats are fascinated by catnip. The reason is that these plants contain a chemical that, when eaten, makes cats 'feel good'. Initially it acts as a stimulant and a cat will chew excitedly, but if she eats spider plant routinely, she becomes a dull feline with reduced physical activity.

25. Sometimes I have this terrible urge. I lie in wait until my human returns. Then I stalk her and attack her ankles. How can I overcome this feeling of aggression?

Cats who stalk their owners are bored, frustrated felines. They need mental and physical stimulation and are almost always housebound animals. With nothing much to do, they sleep most of the day. The odd moth or fly might come their way, but after removing its wings and batting it about, boredom sets in once more.

The only other living things that these cats ever see are humans. Television is repetitive, food is routine and served at the same time each day, but human movement can be exciting. Human feet move erratically, just the way real prey moves. They start, stop, turn, back up, move forward, stay still, disappear around corners, then reappear again. They become the most exciting thing in a boring existence.

With little else to do, some cats stalk feet. Darting behind camouflage, he slinks low to the ground, unblinkingly staring at his prey. As the feet move through the room, the cat quietly stalks forward. Suddenly he launches his attack, sometimes accompanying it with a banshee yell. Grabbing with his forepaws and kicking with his hind legs, he grasps the ankle and sinks his teeth into the flesh. In a flash he is gone.

The only way cats can overcome this form of hunting aggression is to develop an alternative release for their pent-up emotions. People who don't enjoy this behaviour either should wear high protective leather boots or should play more games with their captive moggies.

26. People think I'm human and always want to protect me and baby me. How can I convince them that underneath my beautiful exterior lurks a wild animal?

People find it difficult to think like cats. It requires a great deal of concentration and leads to furrowed brows and blank stares. It is far easier for them to think like people, but this means that most see human qualities in their cat's behaviour. This is perfectly acceptable when it involves determining whether a cat is hungry or thirsty, cold or ill. Cats share a great majority of behaviours with people, but not all of them.

Many cats don't like being babied, mostly because they aren't babies. They are adult animals with adult wants and emotions, but their softness, size and unblinking stare somehow elicits maternal behaviour from many people, even men. People want to hug cats, stroke them, kiss them and care for them as if they were helpless infants. Women in particular often find that their voices go up an octave when they speak to their cats, and catch themselves uttering motherese gibberish, asking, for example, if Fluffy wants 'dindins'.

Because it is in human nature to behave this way, there is little that cats can do to avoid it. However, unexpectedly unsheathing the claws every now and then can serve as a dramatic reminder that although cats are domesticated wild animals, they are wild animals nonetheless.

27. I am housebound. When I see another cat in my garden, I feel this urgent need to empty my bladder, which I do on the nearest wall. Can you explain why I act this way?

Spraying the walls is a sign of anxiety and frustration. Housebound cats are more likely to become frustrated when they find themselves incarcerated indoors while some feline stranger enters what they consider to be their own territory.

An indoor life is a safe one. Housebound cats are less likely to suffer traumatic injuries, pick up contagious diseases and get into fights. They also live longer. But they develop more behavioural problems, and urine spraying is one of the most common.

What a cat really wants to do when he sees another cat on his territory is to march outside and demand that the interloper leaves. Indoor-outdoor cats simply fly through the cat flap and, through intimidation or an outright attack, see off the stranger. But housebound cats find a glass wall prevents them from behaving normally. They might rant and rave, arch their backs, hiss and spit, but the garden visitor continues to investigate the property or, worst of all, actually marks it with urine and faeces. Unable to do anything about it, the indoor cat sprays urine on the nearest wall or even on the window. By doing so he marks his territory as near to the intruder as he can, while at the same time he calms himself down by surrounding himself with familiar scent. All housebound cats, regardless of sex or the lack of it, might behave in this way.

28. My teeth chatter when I see a bird in the garden. Is this for the same reason as when I see another cat?

Almost. Teeth clicking, or chattering, is a sign of frustration but its origins are different.

When hungry cats capture prey, they kill with a powerful and immediate death bite. The cat's teeth have evolved to fit perfectly between the neckbones of their most common prey, the mouse. When a mouse is caught, in a flash the cat slips it around in his mouth until he feels the neck, then he bites down rapidly, severing the spinal cord at the base of the brain and killing the rodent instantly. Domestic cats are inveterate torturers because most of them are, from a behavioural point of view, arrested in their emotional development and still kitten-like in many ways. Cats that kill to survive are humane killers.

When a housebound cat sees a potential meal sitting on the lawn, but can't get near it, he becomes frustrated. He lowers his head, stares intently at his prey and then, either knowingly or not, his teeth move as they would if he had captured the bird. He carries out a death bite but, in the absence of prey, all that happens is his teeth chatter.

29. A neighbour cat keeps coming in my garden and beating me up. What can I do about it?

Retreat. The chances are the neighbouring cat is either younger, tougher and more experienced or his testicles are still intact.

All cats need territories but some need larger territories than others. A neutered female might be happy with a small back garden, while an unneutered female might want three or four to call her own. If she comes across another cat in one of those gardens, and she feels firm enough about taking over that garden, she will beat up the occupant.

A neutered male will also be happy with either a single garden or only a few back gardens to call his own, but intact males are another matter. The unneutered male might want a territory of twenty back gardens as his turf. Each day he patrols through his kingdom and will fight with any cat he feels is out of line. Normally he will only bully other males, although it is not unusual for him to bully females too.

Fighting is something at which cats learn to excel. The more successful fights they have, the more likely they are to fight again. At the opposite end of the spectrum is learned helplessness. Losers soon come to think of themselves as losers. This is fortunate because

 after a few meetings the resident might accept that the bully will always win, and withdraw after some ritual staring and hissing but before damage ensues.

The only alternative a cat

has is to enlist the support of his human. Humans can, theoretically, arrange for their cats to have hormone injections to enhance their territorial vigour, but because humans are bigger than even the largest feline bullies, they are actually the best defence an intimidated cat has.

30. I mind my own business and never interfere in what other cats are doing, so why do they all hiss and spit at me? I feel like a pariah.

A hierarchy develops in any quorum of cats. In a multiple cat household, for example, one cat will come to think of himself as boss cat and, by hissing, spitting, using aggressive body language and boxing, will lord it over the other cats. In most circumstances the others don't squabble much and seldom fight with each other. They kowtow to the boss, but have a relaxed relationship among themselves. Cats don't form rigid pecking orders as, for example, chickens and people do. But there are situations where this relationship breaks down.

One of the most distressing of these situations to people is when one cat becomes ill or debilitated. Rather than offering a helping paw, the rest of the cats often turn against their ill companion. They hound him, preventing him from using his normal resting space or from eating before they have finished their food. Just as frequently, all house cats might turn on a compatriot who has just returned from the cattery, the veterinary hospital or from mating. Uniformly they hiss and spit at his mere presence.

There is a final situation in which constant harassment occurs and that is when a colony of cats decides, without there being any known extenuating circumstances, that one of their members will be treated as a genuine pariah. They hiss, spit and attack; they do not allow their victim to eat, to drink or even to use the litter tray. The only solution in these circumstances is for people to help the pariah to escape.

31. Whenever I'm really frightened, I amaze myself and turn into a vicious tiger. Would it be better if I ran away when I'm frightened?

No, act like a tiger and others will think you are a tiger.

Cats are small and vulnerable. They may be sophisticated rodent-killing machines, but on their own they stand little chance against larger predators. In North Africa, where the domestic cat emerged a few thousand years ago, there are countless enemies: humans, dogs, larger cat species, even snakes. Running will simply provoke a chase-and-capture response from most of them.

When confronted with a bigger, faster and more powerful enemy, a cat's chances are pretty slim if he ends up in paw-to-paw combat. By using clever psychological warfare there is at least a chance of escape and survival. First, the cat maintains direct eye-to-eye contact; almost all species innately understand that only dominant and secure animals will do this. A direct, unblinking, wide-eyed stare is intimidating. In conjunction with this stare, the confronted cat puffs up his body by arching his back and erecting the hair on his neck and back. This suddenly makes him look considerably bigger than he previously appeared to be. Finally, he retracts his lips, revealing his weapons, shrieks his surprisingly loud shriek, and spits. These activities fluster his enemy, who now thinks twice about attacking, calms down and, with luck, walks, trots or slithers away.

32. Why is it that I like some dogs but am scared out of my wits by others?

Cats are less frightened of individual dogs they know than of other members of that species they have never met. The serious limiting condition here is that the kitten's socialisation period is very short, finishing when she is about seven to eight weeks old.

If a cat is to live with dogs, it is best that she meets and plays with them daily while she is still very young. Inevitably this means it is up to the kitten's breeder to make sure that the kitten meets dogs, people or any other species at an early age. If, for example, kittens are raised with mice or rats, they will never kill them.

The curious fact is that cats seem to recognise breeds. Kittens raised with rats will never kill other rats of the same breed but will readily kill other breeds. A similar relationship seems to develop with dogs. If, for example, a cat is raised with a German shepherd dog, it will certainly not be frightened of that German shepherd and might not be frightened of any other German shepherds. But when it is confronted for the first time with, say, a Yorkshire terrier, it might panic.

When cats are raised with dogs, cats remain in charge but are always at risk. A cat might use a dog's forepaws as scratching posts. She might consider the dog a soft warm cushion to lie beside. She might think that a dog's tail is the ultimate plaything. But if she ever runs too quickly, it can provoke an instinctive chase-and-attack response from even the most benign of canines.

33. As I've grown older I've relaxed and become friendlier with people. Is this senility or am I finally growing up?

It is both. A surprisingly large number of pet cats have been rescued from the street by their humans. These cats were raised in the absence of human companionship and are fearful of such big animals. If the human perseveres, however, she can break through this fear and form a bond of affection, but this remains a special and unique bond between them: it doesn't extend to other people. The rescued cat will only relax when his special human is around. If strangers call, he will dart away and hide.

Given time, lots of time, some cats eventually come to realise that humans in general mean no harm. They accept other family members and regular visitors. They don't disappear at the sound of the doorbell and might even jump up on the laps of certain special humans. People might call this growing up.

Senility also has a role to play in feline friendliness. As the years roll by, some cats in their dotage become kittens again. They develop a greater dependency relationship with their humans. They forgo the independence of youth for the security and comfort of being mothered, something that humans, especially female humans, instantly provide. This occurs at the time of life when the cat's brain is actually shrinking in size and nerve pathways are breaking down. It may be called senility, but it is something that some humans wait a lifetime to have happen.

CHAPTER THREE

Training

34. I am in prison. How can I remain mentally alert when I'm never allowed to walk on roofs, sit on fences or climb trees?

Cats are highly adaptable, which is why they have been so successful in colonising all corners of the world. Their adaptability means that they readily accept the restrictions of an indoor existence but then rely upon humans for many of their needs, including mental and physical stimulation.

Good cat furniture is the first requirement. Cats like climbing; curtains are ideal for this. Cats also enjoy climbing onto kitchen work surfaces because they can often discover cached food there.

Toys are another need. Toys should be designed to be either chased or killed. Ping-pong balls are good chasing toys: at the slightest tap of a paw they run away. Fake mice, especially those covered in rabbit fur, are perfect killing toys. The housebound moggie can satisfy his need to attack by furiously biting it.

People should participate in mental stimulation games with cats. Crushed aluminium on the end of a string, balls of wool, feathers or commercially made cat toys all provide games that people can play with and use to stimulate cats. If people are willing to play games, an indoor life can be interesting and active for a cat, yet safe and secure at the same time.

35. Although I have an indoor litter tray, I prefer to empty my bowels in my own garden. I always cover up my droppings but my human doesn't understand. She locks me indoors. What can I do to convince her to let me out again?

Earth is the natural substance that cats use for toileting. Typically, a cat chooses a specific toilet site, digs a small hole with his forepaws, passes his droppings in the hole, then fastidiously covers them up. He always goes back to the same latrine site and only ever leaves his droppings there. When given the choice between using a litter tray covered in a granular substance or a familiar earth toilet, cats prefer the latter.

Regardless of the logical reason for the cat's behaviour, cat droppings in gardens offend most gardeners. People can compromise by doing two things. First, they should bring in soiled earth from the garden and mix it with the cat's litter. By bringing the outdoors indoors they can convince their cat to use his litter tray. At the same time, they should cover all indoor earth in plant pots with aluminium foil or other protection to prevent their house-restricted feline from toileting there.

Once the cat has used the litter tray, the ratio of earth to litter can gradually be changed, and the litter increased until it predominates. At the same time, all vestiges of the former outdoor latrine site should be removed thoroughly. When the cat has been using his indoor litter tray routinely for three weeks, he can then be let outdoors each day, but only after he has emptied his bowels in the indoor tray.

36. Why do I feel this irresistible need to mess in the neighbour's garden?

Cats that mess in neighbours' gardens are really marking their territories. Typically, these cats spray urine on shrubs and fences and don't bury their droppings. Again, typically, they are unneutered tom cats and more often than not they are feral – not owned by people.

Cats use urine and faeces as territory markers. Each day the territory 'owner' patrols all parts of his domain leaving scent markers that tell other cats of his presence. These scents don't intimidate other cats but their freshness tells the cats how recently the 'owner' of the territory has passed through it.

Neutering is the only assured way of eliminating the problem, but because these troublesome cats are often feral it can be difficult to do. Alternatively, people can make their gardens less appealing to cats by creating odours in them that tell the territory marker that the garden already has a powerful occupant. Commercial products that, to the cat at least, smell like cat urine should be used liberally on all affected garden sites. People should remove any plants such as catmint that attract cats and replace them with flowers such as marigolds that produce odours that cats find either competitive or downright offensive. Finally, a big, butch, neutered, litter tray-using tom cat on patrol is probably the best way to keep a garden cat-free.

37. If I must use a litter tray, which type is best and where should it be kept?

Cats naturally prefer privacy when they toilet, so litter trays should be kept away from areas where there is a lot of activity. Because their natural cleanliness means that cats won't toilet near their food, a litter tray should never be placed near food and water bowls.

When possible, the litter tray should be put in a small quiet room where it doesn't create an obstruction and where it can be cleaned easily. Some fastidious cats need their tray cleaned daily, while others get upset if it is cleaned too regularly. The latter feel more secure when they catch the scent of a previous visit.

The type of tray to use varies with how expert a cat is at using it. Most cats are content to use open trays with sides raised high enough so that they don't kick litter out when they cover their discharges. Others prefer enclosures. Covered trays offer privacy, and some even have charcoal filters to eliminate odour. Because each cat naturally creates her own toilet area, ideally there should be a litter tray for each household feline. These can be arranged close to each other, even side by side. By offering the availability of a personal latrine, people reduce the likelihood of the cat messing elsewhere.

38. What type of litter should I use?

People often choose litter according to their own preferences rather than the needs of their cat. The most
important feature of litter is its feel. If a cat doesn't like
the texture of what she is standing on, she won't use it to
mess in. This doesn't necessarily mean that cats prefer
the texture of their litter to be as much like earth or
sand as possible. Most enjoy those feelings underfoot
but are equally happy with other substrates as well.

Chalk, clay, wood shavings, compressed wood chips,
washable beads, even shredded newspaper can all feel
appealing to certain cats. What they don't like is a
sudden change from one type of litter to another. They
don't like perfume either. Some litters are made to
appeal to humans rather than cats. These release
perfumes of pine or lavender when they get wet:
smells that people like but to which cats are, at best,
indifferent.

Some litters contain odour-absorbing substances that
don't emit covering odours themselves; others are biode-
gradable so that they can be flushed down toilets. The
important point for people to remember is that if they
plan to change a cat's litter from one type or even one
brand to another they should do so gradually. Just as a
little soiled litter should accompany a litter tray when a
cat moves from one home to the next (see question 69),
some of the previous litter should always be mixed
with any new type so that the cat scents and feels the
familiar. If people take this precaution, they reduce the
risk of the cat messing elsewhere.

39. I try to aim straight but lately I seem just to miss my litter tray. Do you have any hints on how I can improve my sanitary habits?

Missing the tray is either a tray problem or a cat problem. The sides of the tray might not be high enough, so when the cat urinates he shoots over the edge. The tray might be too dirty. In this situation the cat hovers near the edge of the tray where the litter is cleaner to stand on, and shoots over the edge. The tray might have unpleasant litter. In these circumstances, the cat might balance on the side of the litter tray and shoot over the edge. In all of these instances the stop-gap solution is to cover the surrounding floor with cling film covered with newspaper so that no urine scent is left outside the tray.

The cat can be the source of the problem too; he might need medical attention. Any medical condition that alters the normal use of the urinary system might cause a cat to miss his litter tray. Urinary infections or the build-up of sharp crystals in the urine result in pain, or at least discomfort, when a cat urinates. Some cats will associate this pain with the litter tray. Because they are so hygienic, they still want to use their tray but hover around the edge for a quick getaway. For some curious reason, discomfort from blocked anal glands can have the same effect. Cats feel pain when they defecate, so they do so either near the edge or completely outside their tray, and sometimes they avoid urinating there as well. If this is the case, a trip to the veterinarian should resolve the problem.

40. Although I have a quiet, secluded litter tray filled with delightful litter, I still feel this urgent desire to urinate on the carpet near the front door. How can I stop doing it?

Even in the best of circumstances cats will sometimes fail to use their litter trays. Once medical reasons have been eliminated, it has to be assumed that the problem is psychological.

Frustration at not getting outside can cause the problem. Or the cat might prefer the feel of the carpet underfoot to the feel of the litter. The paramount question is how it can be stopped. Make sure there are no problems with the existing toileting facilities and prevent the cat from using his unpleasant toileting site by cleaning the area as well as possible. Vinegar works nicely, although vodka is better. Avoid anything with ammonia in it, because it smells too much like urine.

Once the area has been cleaned, cover it with furniture or, if that is impossible, with aluminium foil. Cats don't like the feel of foil underfoot. Alternatively, people should start feeding the offender at that very spot. If he eats there he is unlikely to soil there too.

Some cats are completely recalcitrant when it comes to any form of free-range retraining, and for them there is no alternative other than cage training. Using this method, the cat is restricted to a cage in which there is only enough area for a litter tray, bedding and eating area. He is forced to use the litter tray and within a few weeks has been trained to continue to do so.

41. How can I prevent other cats from using my cat flap?

Only with a padlock. If there is a brazen territorial cat in the vicinity and that cat considers both the outdoors and the indoors to be part of his territory, he will readily enter the house to eat food, chase cats or, worse, spray urine.

Uninvited guests are most likely to use the cat flap at night when there is little indoor activity. They are usually frightened of humans and lack the confidence that is necessary to enter a kitchen while people are using it. The only exception is that when giving chase an intruder can chase a resident back through his own cat flap and right through the house, although this usually needs a pretty dynamic adrenalin surge.

Dogs are good at keeping strangers away. Even the most confident of felines will think twice about using a cat flap again when confronted by a Staffordshire bull terrier or even a Labrador on the inside. Magnetic flaps are also worth investigating. These cat doors can only be opened by cats wearing special magnetic activators on their collars. However, do remember that it is possible that these magnets might interfere with a cat's ability to use electromagnetic air waves for navigation. The best way for a cat to prevent others from using his cat flap is to enlist the help of humans. They can use brooms, water pistols and cushions to convince the interloper to leave. They can tape up the cat flap too.

42. Why do I prefer scratching the sofa to the exceedingly expensive scratching post that my people bought for me?

Cats scratch objects mainly for two reasons: to help manicure their nails; and to leave a scent mark.

The feel or texture of a surface is the most important physical attraction it has for a cat. Purpose-built scratching posts might look delightful to a human, but many just don't feel right to cats. Cats like to extend their nails and then be able to drag them vertically down the material. They enjoy a smooth run. Fortunately, most sofas are made of ideal material, so indoor cats don't have to look any further. The problem is that people don't see it quite that way.

People want cats to claw out-of-the-way objects, but that has no appeal for a feline. Another reason for clawing is to leave a visible territory mark. Cats want their scratching posts located prominently – and sofas usually are. Cats like sturdy scratching posts: sofas and wooden table legs almost always meet that requirement. Rather than buying a ready-made post, people should make one themselves using sturdy wood on a solid base. A fence post is often ideal, and part or all of it can be covered in fabric so that nails catch nicely. If the old three-piece suite is being discarded, try chain-sawing an arm off one of the units and donating it to the cat. If a cat still wants to continue clawing everything, amazing as it sounds, soft plastic nail covers are available. These slip over the claws and prevent damage, although most cats feel foolish when they first wear them.

43. What types of activities should my kittens enjoy while they still live with me?

The more mental and physical stimulation kittens experience while they are young, the more well rounded they will be as adults. Kittens should have all their senses stimulated from the time they are only a few weeks old. This early stimulation helps the brain to grow. More interconnections develop between nerve cells, and this leads to the cats having a greater ability to cope with unexpected circumstances.

With help from people, kittens should play games frequently. They should chase strands of wool and bat balls. They should climb obstacle courses to reach food rewards. They should leap onto objects and then off them. They should learn to pull strings with food attached to the end.

If car journeys are to be a feature of the cat's adult life, he should be introduced to vehicles while he is still very young, even before he is seven weeks old. If, however, he is destined to become an apartment dweller, he should not be let outdoors. The great outdoors is natural and exciting. If kittens experience the delights of outdoor life it can be difficult for them to retire to the admittedly more secure but nevertheless tedious and boring confines of a person's home. If a kitten is headed for an indoor life of luxury he should receive instruction on how to remain mentally active and alert.

Equally important are social games. Kittens should frequently meet other cats, dogs, small children, large children, adults, horses, goats or whatever species they are likely to meet when they are older. Most importantly, they should learn to respond to a name. This is done by people saying the kitten's name while holding the food bowl. The kitten eventually meows for the food and is fed. Soon people only need say the name and this elicits a meow. In a similar way kittens can be taught to 'come' using food rewards. People feed something like cat vitamins, shaking the can each time one is given. Soon the kittens will come when they hear the can being shaken. Teaching these responses makes it more likely that if the cats ever get lost and hear a person calling their name they will respond.

CHAPTER FOUR
Sex

44. *Will neutering change my personality?*

Yes it will, but from the viewpoint of humans it will make a cat a more interesting and enjoyable companion.

Neutering has a more profound effect on males than on females. Male cats are more destructive than females. They are more active and travel greater distances. Males get into more territorial demarcation disputes with other cats and fight over sex more frequently. All of this fighting means that male cats get more fight wounds and abscesses. Because so many virus diseases like FIV, feline AIDS, are transmitted in saliva, they also have a much higher incidence of serious and potentially lethal infections. Neutering makes male cats less destructive. It reduces their activity level to that of the female. And although it doesn't eliminate the desire to fight, that desire diminishes. The neutered tom cat is usually content with a smaller territory to patrol and his personality, in many if not most ways, becomes very similar to that of a female.

The female cat's personality is not as dramatically affected by neutering. Females naturally come into season as daylight hours increase and continue having seasons until they become pregnant or until daylight hours start to diminish. When they do so, they undergo profound behaviour changes, becoming more vocal, lascivious and active. During the rest of the year, when they are not in season, they simply go about the business of being cats.

Neutering simply perpetuates and makes permanent this already long phase of out of season behaviour. In essence, neutering eliminates the sex hormone-related

53

variables to a cat's behaviour, which is why neutered males and females have very similar personalities.

When given the opportunity, many cats opt for the easy life. Neutering makes life a lot easier because it eliminates the hassles associated with searching for sex. Some neutered cats become so laid-back they become downright sedentary, even slothful. These cats have a tendency to gain weight, often laying down a healthy reserve in a fat pad that flaps about between the hind legs. Weight gain associated with neutering can have as profound an effect on personality as neutering. Cats who do not want to become dull and boring should watch their diets and eat moderately.

45. How will I know when I'm old enough and ready to mate?

Male cats undergo physical changes when they reach puberty. They develop thicker skin on their necks to protect them in cat fights. Although not as dramatic as a lion's mane, cheek ruffs grow and these make the face look larger. Males grow bigger than females, in some instances very much bigger. A male Maine Coon, for example, might be double the weight of a typical female, reaching ten kilograms in size. Most easy to recognise, however, is the tang of urine. Male sex hormone turns normally banal, uninteresting urine into a blazing garden of pungent aromas. It's so distinctive that even humans with their relatively unsophisticated noses readily detect when a tom cat has sexually matured. Once this has happened, any time is the right time to mate.

Females undergo more dramatic behaviour changes when they are about to ovulate. Previously quiet and reserved cats become more affectionate and demonstrative. They rub their bodies against furniture and humans. They roll about, stretching themselves in sensuous poses. Some moan or wail, clawing themselves across the floor while their hind legs hang so loosely that some people think they have broken them. In a profound display of sexual abandon they present their posteriors to dogs, humans or anything else at hand, holding their tails up and to the side. However, there is never any vaginal discharge as in other mammals because a cat will not actually ovulate until after she has mated successfully.

46. Why do I have this compulsion to caterwaul at night? It wakes all the humans.

Because cats prefer the security of night, it is the most frequent time for them either to go looking for sex or to have it. Both events can be noisy.

When toms fight over territory, or have duels to see who mates with the receptive female, they can be surprisingly, loudly, fearsomely vocal. Considering their small size, cats produce loud shrieks and growls. Sleeping people seldom hear the preceding hisses, spits and growls, but when two toms enter combat their high-pitched shrieks and war cries become the loudest noises of the night.

Mating is a noisy business too. The receptive female initially but quietly spurns the male's advances, but once she is ready, she raises her haunches, he stands astride and enters her. Mating occurs quite quickly, but withdrawal creates a problem because the male's penis is covered with backward-pointing fish hook-like barbs. These cause no trauma or discomfort as he enters, but when he withdraws they lacerate the lining of the vagina and the female suddenly shrieks, once more waking up all nearby humans. She turns to bite him, but cleverly he has bitten into her neck and this prevents her from doing so. When her shriek stops, he dismounts and they both wash their private parts.

47. If it hurts so much that I scream, why am I so willing to have sex again a few minutes later?

Most mammals, apart from cats – dogs, cows, goats, pigs, sheep, horses – in fact all domestic animals, follow a similar sex cycle. The female's ovaries produce eggs. Only when these are released from the ovaries into the fallopian tubes are the females ready and willing to mate with the opportunist male who is always ready, willing and able.

Cats need to mate *before* they release their eggs. Mating is the actual event that stimulates ovulation. The problem is that one mating is usually not a sufficient stimulus to get the system working. Humans assume that because the female shrieks in apparent pain and tries to bite the male she finds sex painful and therefore unpleasant. Human value judgements do not apply to animal behaviour, however, because after she has groomed herself, the female stands more willingly and mates again.

Once more she will shriek as the male withdraws his penis, but she becomes more demonstrative and actively solicits further matings. If the male claims exhaustion or a headache, she presents herself to one of the other males patiently waiting his turn. As the number of matings increases so, too, does the likelihood that she will ovulate successfully.

48. How should I take care of myself while I'm pregnant?

Pregnant cats can carry on routinely until just before birth. At first the only sign of pregnancy occurs when the nipples become pinker and larger. There are no other outward signs, although people might feel small golf ball-sized masses in the abdomen about three weeks after conception.

Hormone changes cause the cat to become quieter. She will roam less, rest more and eventually start eating more. During pregnancy she has to maintain her own body but also build the bodies of her kittens. To do so she needs a normal well-balanced diet during the first few weeks, but as the pregnancy continues her nutritional demands increase. Minerals such as iron and calcium are needed; high quality protein is too. Pregnant cats remain agile and swift, so they have little difficulty in capturing nutritious meals.

It is only as the pregnancy enters its second and final month that a cat shows the first visible change in her body. Her abdomen increases in size, and as it does so her centre of gravity changes. This makes climbing and jumping more treacherous. Just before birth she carries the equivalent of an extra kilogram of weight in her belly. She should rest as much as possible and should be allowed to select a nest where she can produce her litter safely and contentedly.

49. When I was in season recently I had sex with five tom cats. Which one is the father of my new litter of kittens?

No one knows. Promiscuity in cats is natural and possibly quite useful.

Take lions for example. Under normal circumstances the lion pride consists of a group of females and only one male, who has sole mating rights with his harem. But when that male is ultimately defeated and displaced by another male, the new one sets about ruthlessly killing the cubs sired by his predecessor. Infanticide means that all of the young will henceforth carry the genes of the new king.

When allowed to do so, and that means when a good source of food is readily available, free-living domestic cats create a society that is in many ways similar to a lion pride. A group of females lives communally. The only males allowed to live with them are their own adolescent male kittens who have not yet left home.

There may be several tom cats in the vicinity, any one of which might be dominant at a given time. The female mates with any or all of them, making the fatherhood of the kittens difficult or even impossible to determine. By doing so, the queen unwittingly but dramatically reduces the likelihood of subsequent male infant killing.

50. Will my kittens' father help me to raise my litter?

Don't expect him to. Tom cats rarely show any interest in their kittens.

Some toms might hang around and, in a rather banal way, watch the birth. Even more rarely, some toms might give the odd lick, but this is the exception not the rule. Cats are old-fashioned feminists, so rather than counting on help from males, they help each other.

When cats live in natural colonies and that means when they, rather than humans, decide who should stay and who should leave, they form matriarchies, colonies of genetically related females. The colony might consist of a great-grandmother, a couple of grandmothers, several mothers and several litters of kittens. Other colony members might assist with the birth, even to the extent of chewing off umbilical cords when a naïve and inexperienced mother fails to do so.

After birth, other females sometimes help clean the kittens and continue to do so as they grow. If a mother goes hunting, other lactating mothers feed her kittens; if she doesn't want to go hunting, they sometimes bring food back for her. As her kittens grow, other colony members will help to teach them how to hunt. Although the natural mother is always the best teacher, other females might bring back still live food and give it to the litter so that they can learn to kill. Matriarchal colony life means that females can be released from the constant demands of motherhood without having to rely upon unwilling tom cats.

51. Why do I prefer to make love to cushions rather than to other cats?

Probably because that's all there is. Sexual behaviour in males does not necessarily rely upon either attaining puberty or experience. The female cat's brain remains sexually 'neutral' until she reaches puberty, but tom cats, and most other male mammals for that matter, have brains that are 'male' from birth.

What happens is that just before birth the cat's testicles produce male hormone, this circulates in his body and actually influences the development of his brain. This means that certain behaviours that are 'masculine', like pelvic thrusting, are there from birth: kittens, for instance, might occasionally pelvic thrust on each other. Adult toms will use this action as a form of dominance as well as when mating.

Naturally, if the desire is there and given the opportunity, a tom cat will pelvic thrust on another cat. But in the absence of another cat, even in the absence of his testicles, he might still carry out this function on whatever is available. Some cats will use their human's arm or leg, but these objects often refuse to stay still, so other cats prefer less troublesome articles.

52. I like sleeping with strangers, especially humans, but what is it about their armpits that thrills me so?

Many cats come to enjoy the warmth and comfort of human companionship, and so choose to sleep on or even in bed with people.

When cats behave this way they are really behaving as kittens, seeking out the satisfying security of a mother. They nestle on their human, contentedly curling up in a more relaxed way than they would ever behave with another cat. This close proximity brings them in contact with human odours, and armpits are particularly appealing.

Odour is very important to cats. Immediately after birth the mother licks her kittens and her own nipples, laying down an odour trail for her kittens to follow to find food. Each nipple is surrounded with glands that produce subtle differences in odour, too subtle for humans to scent but strong enough for kittens to recognise the different smells and always go back to their own personal nipple.

Human armpits are riddled with odour-producing glands, some of which are sex related and become active at puberty. One or more of these odours is particularly appealing to cats, and is so close to the odour of their mother's nipples that it stimulates the now adult animals to nestle, tread with their forepaws to stimulate milk flow, and suck. If humans don't like this activity, they should wash their armpits more frequently and use deodorant.

53. Why do I have this compulsion to suck people's woollies and why do I dribble when I do it?

Wool sucking occurs when kittens are weaned from their mothers too soon. Although a kitten needs his mother's milk for only the first six weeks of his life, sucking continues for another variable period of six to ten weeks, and changes from a nutritional to a social behaviour. People think that because their mother's milk is no longer necessary for their survival, kittens are ready to leave their mothers. Litters are often disbanded when the kittens are only eight to ten weeks old. This is too soon.

Suckling is a comfort behaviour which naturally ends when the kitten becomes emotionally mature. If it artificially ends through separation, that kitten will continue to need to suckle just as humans need to suck thumbs. The soft texture of wool is similar to a mother's fur, which is why they become wool suckers.

In most cats wool sucking is a behaviour they carry with them from kittenhood, but there is a curious anomaly: some Siamese cats start wool sucking only *after* they reach puberty. These cats have had normal weaning, they don't indulge in comfort behaviour, but after they have matured they develop a passion for mohair. This is a unique genetic form of wool sucking that occurs in some lines of Siamese and Burmese cats.

CHAPTER FIVE
Diet

54. My people don't eat any meat and, because they treat me as part of their family, they don't want me to eat meat either. Can I survive on a vegetarian diet?

No, you can't. People, and dogs and pigs, are omnivores. They have digestion systems and biochemical processes that convert substances found in vegetable matter into all of the essentials for life. But cats are true carnivores. Their bodies cannot manufacture certain fats and amino acids that are necessary for survival. The only way they can acquire these nutrients is by eating the bodies of animals that have already manufactured them.

Throughout the world cats have varied diets: in North America they enjoy chipmunks; in Europe, mice and rabbits; in Australia, ring tail possums and, on isolated southerly islands, penguins and noddies. These staples are supplemented with other living delicacies: grasshoppers, spiders, moths and flies. Cats are rather indifferent to sweet flavours, preferring their food to be slightly salty and served at body temperature.

People can create disease problems by feeding cats only one type of meat. Variety is necessary; cats get bored with the same food day after day. They willingly try something new, but if they don't like it, they will refuse it the next time it is served. Whatever it is, it must be made from the bodies of other animals.

55. *Will I choke on bones if I try to eat them?*

The only bones to avoid are fishbones. They can stick in the back of the throat and cause a cat to choke.

Many books that cat people read say that cats should never be fed bones. Cats, of course, pay no attention to this silly advice. They never leave tidy piles of mouse and bird skeletons after a feast. Although most cats will pluck a feathered meal before consumption, and some will make abdominal incisions in mice, removing the gall bladder and intestines before eating, all the bones get swallowed.

Eating habits are learned at an early age. If a kitten is routinely fed bones, she will mature into an adult bone-eater. At first she will bat a chicken wing around as if it were prey, then pounce on it, bite it, dance around it and finally eat it.

Chewing on bones is the way cats keep their teeth and gums healthy and clean. Unlike dogs, who bolt down their food, most cats eat almost gracefully, carefully chewing bone to small pieces. In the wild small rodents are often swallowed whole, and bones are passed by cats in much the same way that owls pass the bones of their prey. Nutritionally affluent pet cats are more likely to chew the meat off the bone, perhaps eating the cartilage but leaving the difficult bits. Other than when chicken bones get stuck on cats' teeth, veterinarians seldom see medical emergencies caused by eating skeletons.

56. I love the crunch of dry cat food: it's just like munching on succulent mouse bones. However, I've heard that eating it causes urinary problems. Are dry foods safe to eat?

In most instances they are, but there are some circumstances in which they should be avoided.

Just like humans, mice and birds are about eighty per cent liquid. When cats eat these natural prey, they get most of the daily liquid they need. That's why some cats seldom drink anything. Canned cat food also contains about eighty per cent liquid, so when cats eat that they also satisfy their fluids requirements, but soft moist food is only thirty per cent moisture and dry food is considerably less. When these foods are eaten, cats simply *must* drink more fluids to meet their body's needs.

Drinking enough fluids is not a problem when cats eat dry food from kittenhood. They learn as youngsters to drink extra water or milk, which people should always provide. Once they have matured, however, some cats get very settled in their ways. They might like or even positively crave the taste of dry food but refuse to increase their water consumption. When this happens, their urine becomes more concentrated. Any crystals that might form in natural urine also concentrate, and this can lead to inflammation of the bladder wall and subsequent infection.

Dry food is safe to feed young cats, but if mature cats switch to it people should watch carefully to make sure that their fluid intake is high enough.

57. *Skimmed milk is quite tasty, homogenised milk is delicious, and cream is heaven. My problem is that my owners think that drinking it gives me diarrhoea and worms. How can this be true?*

Natural cat mother's milk is very rich, concentrated and has a much higher fat content than cow's milk. In fact, it is closer to cream than to skimmed milk in consistency. Cats are opportunist feeders, and for many a bowl of milk is a deliciously nutritious reminder of mummy.

When cats are still kittens living on their mother's milk they have the chemical ability to break down the sugar in milk, lactose, and use it for energy. As they mature, some cats lose the ability to break down lactose; when it is not broken down, its presence in the intestines causes diarrhoea. This is why dairy products sometimes cause diarrhoea in some cats.

If a cat enjoys dairy products, she can either be fed foods like yoghurt or soured cream, in which lactose has already been broken down, or special 'cat's milk', a lactose-free product made essentially for cats.

Drinking milk is no more likely to cause worms in cats than it is in humans.

58. Each winter people give their children extra vitamins and minerals. Should I take some too?

Vitamins, minerals and fatty acids are necessary to maintain good health and prevent disease. Too few or too many can cause problems. Although some vitamins are manufactured by bacteria in the intestines, most vitamins and minerals are consumed in food.

When a cat eats a natural diet, he gets all the vitamins and minerals he needs. As long as a good healthy diet is fed throughout the winter, there is no need to supplement it. A potential problem can occur when cats eat commercially produced cat food. It was discovered only recently that a type of heart disease that veterinarians were seeing in cats with increasing frequency was caused by a taurine deficiency in their canned food. So, if a cat eats the same commercial food each day, either his diet should be varied occasionally, just to be safe, or he should consume a general vitamin and mineral supplement.

Fortunately, many supplements are actually quite tasty. Some cats will scale work surfaces and kitchen cupboards just to be near their supplements. If this is the case, the supplement can serve another purpose. People can use vitamin tablets as bribes, training rewards to be given when a cat meows when his name is called or comes back on command.

59. I am a grazer. Although I sometimes wolf down my meal at a single sitting, I prefer to come back frequently and eat little bits at a time. Do I eat this way because I am bored?

When food is plentiful, grazing is a normal way of eating. If cats are raised with food constantly available, they will eat small portions at a time throughout the day. Typically a cat will snack every couple of hours.

This is not the behaviour of the lone hunter. When a cat has to kill to survive, she will fill her stomach when hungry and store bodies in a larder for when food is in short supply.

People feed cats in two completely different ways. Feeding small meals several times a day trains a cat to lick the bowl clean at certain times. This is how most people feed cats, which is why house cats usually polish off the plate as soon as it is filled. Other people, especially those who feed dry cat foods, fill the bowl to excess once daily. This trains the cat to pick at her food because there is always more than enough to satisfy immediate hunger.

There is no right or wrong way to feed a feline. Most cats are sensible about what they eat, which is why there are far fewer fat cats than fat dogs or fat people.

60. I am one of those fat cats. I am quiet, sedentary, slothful, and happy the way I am. Sex doesn't interest me after I had an operation when I was young. I weigh twice as much as other cats with my bone structure. Honest. Is it really harmful to be overweight?

Only ten per cent of cats are overweight, far fewer than the thirty per cent of dogs. And life expectancy does not seem to be shortened significantly for these overweight cats. But there are other problems.

The cat most likely to run to fat is neutered, housebound, naturally lazy and rather indolent. Housebound cats live much longer than their outdoor cousins. Similarly, neutered cats have longer lives than intact brethren, so although fat cats live long lives, there are other factors that also help to achieve longevity.

On the negative side, fat lazy cats have a far higher incidence of urinary problems than lean ones. They are less agile and injure their joints more frequently. The 'Catch 22' of their lives is that as they get fatter, and therefore less active, several risk factors actually diminish. The consequence is that if a cat is happy and content to be overweight, that condition is not overwhelmingly harmful. Humans have a mania about fatness. In their culture fat is a negative. Because they treat cats as members of their human family, they apply the 'fat is bad' rule to their cats too. For a housebound feline it ain't necessarily so.

61. Although I am housebound, I constantly crave food and never have enough, but I am as skinny as a rake. My nerves are on edge and I jump at the slightest noise. Should I be concerned?

Yes. If a cat eats well but loses weight, there is something seriously wrong.

When the kidneys fail to filter properly, the protein in food is lost in the urine. Cats with kidney failure eat well and lose weight, but they almost always drink more than they previously did. When kidneys don't work well, a cat should eat food with less protein.

If certain problems develop in the intestines, a cat can lose his ability to absorb food into his body. In these instances, although he is eating copious amounts of food, the cat is unable to digest the nourishment it contains and is actually starving. Treatments will vary according to the specific cause of 'malabsorption'.

In this instance, however, skinniness is probably caused by an overactive thyroid gland, a problem of dramatically increasing frequency. When too much thyroid hormone circulates in the body, metabolic activity increases. There is no increase in thirst but the heart rate soars, and cats become, well, 'nervy'. Their high energy requirement means that they need massive amounts of food daily. Fortunately, the problem can be diagnosed accurately with blood tests, and treated either by removing the offending thyroid gland or by taking tablets that suppress the overactive gland.

62. My diet hasn't changed and I've never been much of a drinker, but lately I've had this insatiable thirst. Should I be concerned?

Any sudden increase in thirst is a cause for concern. Sometimes the cause is obvious, like changing from moist to dry food, but when there are no known reasons, people should assume that something is wrong.

Poor functioning of the kidney causes an increased thirst that is accompanied initially by increased hunger. Although cats don't like it, a diet change is mandatory, either to one of the commercially produced low-protein diets for cats with kidney problems or to a home-made diet of, for example, one part liver to five parts rice.

Diabetes, more common in cats than many people realise, will also show itself through an increased thirst. Diabetic cats pass sugar in their urine, which makes diagnosis quite easy. People can help diabetic cats by injecting them daily with insulin. Some people find the prospect of giving injections quite intimidating but soon learn that it is a simple and painless procedure.

Sometimes cats lose the ability to concentrate their urine. It passes out in a dilute form, so dilute that this problem, too, can be diagnosed by checking its concentration.

Treatment, curiously, is with daily eyedrops. Other conditions, such as bladder infections or the side effects of drugs, can also increase thirst. Whatever the reason, people should always contact their veterinarian for advice.

63. Why do I prefer drinking water from a dripping tap rather than from the water bowl that people kindly fill for me?

Simply because it is fun. Some people like to do the same, only they use the garden hose. Dogs do too. They enjoy the refreshing feeling of biting into water shooting from a hose. (Some big dogs prefer drinking from toilet bowls.)

Cats drink by cupping their tongues into a classical spoon shape and lapping. Taps drip perfect portions of water to fill the cup – no mess, no waste, aesthetically pleasing. Finding his own source of water satisfies a cat's need for mental stimulation. Rather than drinking from a boring bowl marked 'CAT', he leaps up, treads warily among soap and utensils, then leans forward and drinks from his own natural waterfall.

Conscientious humans often intentionally leave taps dripping, and this human behaviour reinforces the cat's behaviour. If a cat learns that dripping water is always available, what starts out as a curiosity soon becomes a habit.

Finally, unlike many other species of mammals, cats can taste water. They have special taste buds for doing so. It could simply be that fresh running tap water tastes better than water that has been sitting in a bowl for several hours.

64. Sometimes I get this unpleasant feeling in my gut, start heaving, and vomit up a massive hairball. Is there anything I can do to prevent this happening?

Vomiting hairballs is perfectly normal, but there are several things cats and people can do to reduce the frequency of these incidents.

A cat's fur naturally grows longer in winter and becomes shorter in summer. This means that hair grows and is moulted seasonally. Outdoor cats are more prone to develop hairballs in the spring when they groom away and swallow their winter coat. Indoor cats are more likely to grow and moult hair year round. They, too, groom themselves by licking and swallowing hair. Finally, long-haired cats have 'unnatural' coats. When this hair is swallowed, it is more likely to form a ball of fur in the stomach.

People should brush their cats routinely to rid them of all dead and excess hair. But, additionally, during the moulting season, they should feed their cats one of the many substances that help them to pass the hairball through their intestines and out the usual way.

65. People were surprised when I recently tried to eat the guinea pig they brought home. Is there anything wrong if I did?

Nutritionally there is nothing wrong, but morally it creates a dilemma for humans. In one set of circumstances an animal is an acceptable source of nourishment, but under another, the animal develops an identity and so claims a degree of protection from humans.

Keeping an animal as a pet or as a companion alters the moral value of that particular animal and sometimes its entire species. Of course the guinea pig is an excellent source of nourishment for people, as well as for cats, in some parts of the world. Some people eat horse and others find it offensive. Some people eat dog and others find that an abomination.

Because a cat is treated as a member of the family, people sometimes expect him to live by the moral rules that they set for themselves and their children. This can be ludicrous, because the rules simply don't apply. These are the circumstances in which humanising cats goes too far. People should not let a cat eat any other animal kept as a pet, but they should be neither surprised nor upset when a cat lives by his own biological principles and tries to do so.

CHAPTER SIX

Travel

66. *My humans are moving. Should I go with them?*

Probably. Because they are not pack animals, cats don't form as dependent a relationship with people as dogs do. There are exceptions of course. The Oriental cats in particular – Siamese, Burmese, Tonkinese – are more likely to develop strong attachments to people than long-haired or Persian cats.

Instead, cats form attachments to their territory. (The man who tried to avoid taxes by selling his cat for a considerable sum of money, and throwing in his house for free, was making a valid statement about cat behaviour.) This means that if a cat is independent and secure where she lives, and the people moving into her home are cataholics, willing to care for a non-paying guest, then there is no cat reason why she should not stay where she is.

There are human reasons for moving on with people. Humans usually feel remorseless guilt at leaving a cat behind, and that is only proper. Caring for a cat carries with it the moral responsibility for ensuring that she remains safe, healthy and actively enjoys life. Most cats should give in to these human moral needs and graciously pack up their litter trays and baskets when their people pack up their own.

67. I've heard stories that I have ESP and can use it to find my way home when I get lost. Is this true?

The unfortunate fact is that the vast majority of lost cats never find their way home. Yet stories persist of cats finding their way home over mountains, rivers and highways, and these stories are common throughout the world. There are two possible explanations.

It is likely that cats do have a form of ESP and can use electromagnetic waves to orientate themselves. Certain earthquakes, for example, are preceded by electromagnetic changes in the atmosphere; in the minutes before the earthquake, certain cats can become excited or agitated. It seems they can sense the electromagnetic change. Other experiments show that some cats can orientate towards home as long as they are not too far from it, but wearing a magnet on the collar interferes with this orientation.

This doesn't answer the question of how cats can find their way home over hundreds of kilometres. Part of the answer is that people want to believe these stories. It is part of folklore that, against all odds, a loyal cat will find her way back to the security and protection of her human home. It is a story that people want to believe so desperately that honest mistakes are made. For instance, they can mistake another cat for the one they left behind several years and several hundred miles back. Although it is always possible that the occasional story is true, the stories in general are, regrettably, often based upon human fantasy.

79

68. Is there anything I should do to get ready for a car journey?

Ideally cats should empty their bowels and bladders before car journeys, but unfortunately, just like kids, many forget. If a cat previously has suffered from car sickness, he should not eat for eight to twelve hours before the trip. But even without eating, some cats with delicate stomachs will still dribble at the mouth because of motion sickness. These cats should take anti-nausea tablets before they leave.

Cats should usually stay in cat baskets while in the car: in their panic some try to use their paws as blind-folds on the driver. On long car rides the driver should stop every few hours to allow the cat out of the basket and onto a litter tray just in case he needs to evacuate his bowels. Others can be given a little fresh air if they are accustomed to walking on a lead.

Depending on personality, some cats want to see everything that is happening; others want to hide. The cat basket should be placed in an appropriate location. Regardless of location, however, some felines become furiously vocal in cars. They mutter non-stop expletives. People often want to drug these cats, and there are some extreme circumstances when drugging is necessary. It is usually best to avoid any medications, because the ones that are powerful enough to stop the noise often last much longer than a typical car journey.

69. Are there any tips on how I can best settle into my new home?

Take used litter with you, stick to a single room for a few days, and don't go outside for several weeks.

Everything in a new home smells different and foreign, and therefore a bit frightening. But a cat who smells her own soiled litter instantly recognises something refreshingly familiar. Moving time is the wrong time for people to change the type of litter a cat uses. Instead, they should take the familiar tray and a sample of used litter to provide for their travel-shocked feline.

By staying in one room, a cat will find new secure hiding and sleeping places. With the door closed, she is safe from the mayhem going on elsewhere as unpacking continues. Once human activity has settled down, the cat can emerge and investigate her new territory.

During the next two to three weeks a sensible cat will investigate every single corner of the new house. She will find observation platforms and hiding places, and will take over areas to be used as resting stations. Only when she is fully relaxed and comfortable at home should she be let outdoors for the first time. She should be let out initially on a long lead so that she doesn't run away. And, of course, any sensible cat will wear an identification tag with her new address and phone number. Finally, she should be let out on her own, without a lead, but on an empty stomach, so that she is likely to nip back home soon for a meal.

70. How can I make sure I don't get lost?

Don't go outside. If cats do go out, there is always the risk of getting in a muddle and not finding the way back home.

All cats should wear permanent identification. They can be tattooed, they can be implanted with silicon microchips or they can wear name tags. However, there is always the possibility that a name tag will get lost when an elasticated collar gets caught on something.

Sensible people train their cats to meow when they hear their name called. This helps when a cat gets lost and people are out looking for him. They also take pictures of their cat and mark down distinguishing characteristics that can be used on posters if he goes missing.

The more time a cat spends on his territory, the less likely he is to get lost. Cats that venture out for the first time are most at risk, and should be watched carefully. Once their cat's routine is understood, people will have a slightly better idea of where to look when he does not return home.

Regardless of how well he knows his territory, every outdoor cat should wear a collar. People worry that the collar will catch on objects and the cat will be unable to free himself. This is a minor risk compared to that run by carrying no identification at all.

71. I've heard stories that the most dangerous thing I will ever meet is a road. Is that really true?

Unfortunately, it is. Cars are cats' most lethal enemy, more dangerous than dogs, humans or even infectious diseases.

Road sense isn't part of a cat's learning curriculum. In their natural evolution, there is nothing that prepares them for the speed and size of cars. No animal's eyes are as terrifying to cats as the headlight eyes of a car swiftly bearing down on them. It is a sight that freezes cats to the road and leads to their deaths by the thousand.

People can try to protect cats by making sure they wear reflective collars. Even some flea collars come with reflective bands, which help drivers to see them. The only guaranteed safety measure is to prevent cats from crossing roads, in most instances an impossibility. People must make a basic decision: whether or not to let their cats outdoors. If cats stay indoors, they will be healthier and will live longer. If they go outdoors, they will use their bodies and their minds in ways that nature intended them to be used. Some people have the good fortune to find a happy compromise, where cats have access to the natural world of a back garden but are prevented from gaining access to lethal highways.

72. My people are going on holiday and tell me I'm going to kitty camp for three weeks. What is it like?

A boarding cattery can be traumatic for some cats, and a bland experience for others. Once more, the cat's reaction will vary according to her personality. Introvert cats find the experience unpleasant. They shrink back in their chalets, go on wildcat strike, and refuse to be touched or to eat. Extravert cats look upon boarding as a learning experience in which they meet new and interesting people, taste new foods and smell new cat smells.

Boarding catteries are equipped in a variety of styles. The conscientious cat person chooses one with full amenities so that on good days a cat can be gregariously active and on rotten days she can curl up in a corner and sulk. Individual heated chalets with outdoor runs are best. In these facilities a cat can sit outside and commune with nature or curl up by the fire and doze until her owner returns.

The best catteries cater to an individual's dietary requirements by providing a full menu of different commercial diets or by asking the cat's human to bring special food. If a cat has a favourite basket and toy, these should travel with her while her owner is away.

73. I've been given the option of kitty camp or living on my own with a neighbour coming in to feed me. What is best for me?

Both options have their strong points. Good catteries are professionally run and do everything possible to make sure that a cat is content, eats well and is groomed during his stay. The owners want to ensure that the cat will come back the next time his people go away.

One of the drawbacks to catteries is that whenever there is a concentration of cats the risk of infectious disease increases. Many of these diseases are spread through saliva or by sneezing, and although cats might never come in contact with each other in the cattery, it is still possible for staff unwittingly to spread infections as they go from one cat to another.

The advantage of having a neighbour come in to feed a cat is obvious. The cat remains on his own territory and does not suffer the potential emotional distress of being removed from both his people and his territory.

Feeding a cat and cleaning his litter tray is not enough, however. If a cat who is used to human company is left alone all day, he quite simply gets bored. The best arrangement is to find a saintly neighbour who is willing to come in each day to play with the cat as well as to feed and groom him.

74. Now they tell me that we're all going to move to another country and that I'll be placed in quarantine when I arrive. Is that a fair thing to do to me?

The period of quarantine varies from a few weeks to six months depending upon which country is being entered. It exists in case a cat is carrying and develops rabies. Quarantine catteries are much the same as boarding catteries, so if a cat has previously been a boarder and has not minded the experience, quarantine will be no worse.

Whether quarantine is really necessary any longer is another matter. It seems ludicrous that cats from Britain or Australia or New Zealand or any other rabies-free country should have to be quarantined under rabies prevention rules anywhere. What is more controversial are suggestions that, with new vaccines and methods of cat identification, there are now ways to abolish quarantine and still protect rabies-free countries from the scourge of the disease. One plausible suggestion is that cats carry passports that confirm they have been vaccinated, blood-tested and found to have good protection against rabies. Their passports would also contain their personal silicon microchip number, which could be confirmed by sweeping a reader over the cat's shoulders and checking that the implanted chip number is the same as the number on the passport. Under this type of arrangement people could take cats more freely from one country to another without the worry of quarantine.

CHAPTER SEVEN

Illness and Disease

75. I don't know why, but whenever I don't feel well I want to hide. Isn't there a better way for me to tell my humans I'm unwell?

Hiding is only natural. Because of their size, ill cats are easy pickings for larger predators, so whenever they don't feel completely fit, their self-preservation drive tells them to tuck themselves away until they feel better.

This is beneficial in other ways too. A quiet, retired cat uses up less energy, and therefore needs less to eat. In fact, not eating is sometimes a positive advantage when a cat has a fever. Certain life-saving chemical changes occur more effectively when food is not being consumed.

Hiding does create a problem for people, however. Just as cats don't wear their emotions on their sleeves, the fact that they don't go running to mummy when they injure themselves means that people are sometimes unaware that a cat is seriously ill until a disease has firmly taken root. This means that cat people have to be more vigilant than dog people. They should be suspicious of any changes in their cat's routine, changes such as eating, drinking, sleeping or exercising more or less. They should monitor the state of the cat's litter tray, and report any changes either in how a cat evacuates or the consistency of the products to a veterinarian. They should also make the simple investment of a once-a-year physical examination of the cat.

76. When people aren't watching, I enjoy licking my back and biting my fur, but I do it so much I leave scabs and broken hair. How can I stop mutilating myself?

The ideal way for a cat to stop mutilating herself is to determine the cause of the problem and then to eliminate it. By far the most likely cause of this form of mutilation is an allergic reaction to flea saliva.

For a flea to suck out its meal successfully, it must inject an anticoagulant into the cat. Many cats are so allergic to this saliva that they bite and chew themselves until their backs are covered in scabs. All that is needed is a single flea. Other housemate cats who do not suffer from flea saliva allergy will maintain perfectly normal coats. To stop this form of mutilation, the house and all its livestock must be treated for fleas.

Sometimes, a hormonal upset can cause cats to mutilate themselves. At least this is what some veterinarians feel, because the condition responds when the cat is put on a hormone supplement. It is very possible, however, that the supplement acts like cortisone and simply reduces the itch rather than really treats the condition.

Surprisingly, food allergies can also cause cats to mutilate themselves. This only occurs after a cat has been eating the same food for a very considerable time. To find out whether food is the cause, a cat should change her diet for at the very least four weeks to something she has either never or infrequently eaten before. If the self-mutilating diminishes, it can be inferred that food has been the cause of the problem.

77. Sometimes I lick my hind legs and belly, swallowing all the fur and leaving my under-carriage almost bald. Are the causes the same as when I chew my back?

Sometimes, but there is another more important one. Cats have anal sacs located on either side of the anal opening, which discharge their perfume each time the cat passes droppings. The perfume acts as a daily news-paper for other cats, telling them where and even when the manufacturer of the perfume was last in the area, whether the cat was a male or female, and if female, whether she is in season.

Both with advancing age and with a restriction of territory, these anal glands are underused. The liquid substance within becomes drier, eventually changing to the consistency of wet sand. When this happens the cat finds it difficult or even impossible to discharge the glands and they become impacted, feeling like two small peas on either side of the anal opening. The sensible cat tries manually to empty the glands, and in the absence of forefingers and thumbs, he licks.

He licks around the anal region but gets carried away, grooming the thighs and the abdomen. After a week of licking, a curiously bald undercarriage emerges. When this happens cats need help from people. The anal glands need to be squeezed empty, and sometimes will benefit from being syringed clean with an oily lubricat-ing antibiotic solution. At the same time it is sometimes sensible for the cat to receive an injection of cortisone, an anti-inflammatory, to break the cycle of licking.

78. I am not itchy and I don't pull my hair out. Why then do I have such an embarrassing case of dandruff?

Dandruff appears when a cat is unwell, when the skin is infested with mites, or simply when humidity is low and the central heating is on full blast.

Any under-the-weather cat is likely to have a scurfy coat. Just visiting the veterinary clinic is enough to make some coats look slightly dull, but they shine up again as soon as the cat leaves the clinic and is 'out of danger'.

A more likely cause of dandruff, especially in young cats and those that frequently meet others, is a small mite that burrows in the debris on the surface of the skin. This mite seldom causes a great deal of itchiness, although curiously it is transmissible to human skin, where it can cause a short-lived itch. Insecticidal shampoos are very effective; as the mites are destroyed, the accompanying dandruff disappears.

Each winter, when central heating goes on, the condition of the house cat's coat deteriorates. Generally speaking, the higher the temperature and the lower the relative humidity, the more dandruffy will be the cat's coat. People who enjoy keeping their central heating at a high temperature should consider installing a humidifier for both their and their cat's benefit.

79. My people have discovered from their doctor that they have ringworm, and I have been accused of giving it to them. How can that be when I don't have the disease and my coat is in pristine condition?

Ringworm is a fungus disease transmitted by microscopic spores. The name is unfortunate, because it has nothing in common with worms and is more comparable to athlete's foot in humans.

Soil is the mother lode for this disease. Spores can survive in the soil for years, waiting to be picked up by animal skin, which becomes infected, and usually shows ring-shaped sores. Cats, especially long-haired cats, can act as symptomless carriers. The cat picks up spores that travel home in her fur. That fur is then shed at home or is petted by humans or is rubbed against humans, and the spores are transmitted without the cat suffering from the disease.

When people have ringworm, the cats and the home should also be treated. The environment needs to be thoroughly cleaned of all cat hair. If it is possible, carpets should be given a wash with an anti-fungal solution, so should cats. They should be checked by a veterinarian to make sure their nails are not infected. If they are, the cat needs an antibiotic called griseofulvin, given by mouth for several weeks. Other sore spots are treated with the same cream or ointment that people use, and the cat should be sponged down several times with an anti-fungal solution. Whenever possible, the mother lode of spores should be treated as well.

80. Are there other diseases I can unwittingly pass on to people even though I'm healthy?

People are, naturally, much more likely to pick up diseases from other people than they ever are from cats, but there are several diseases or infestations in addition to fleas, mange and ringworm that cats can transmit to humans.

Rabies is the most serious, and in all affected areas of the world cats should be vaccinated against it. A cat bite can carry a variety of unpleasant germs, including tetanus bacteria and another that causes cat scratch fever, although this is extremely uncommon. Cat bites should be disinfected immediately, and if they are deep, the person should be treated with tetanus antitoxin.

Potentially the most common serious transmissible disease is toxoplasmosis. A cat eats an animal that carries toxoplasma. The toxo organism multiplies in the cat's body, but unlike other animals, cats pass infective stages of toxo in their faeces. People can then pick up the disease by handling contaminated cat droppings.

Cats will only pass toxo in their faeces for about three weeks. If they pass it into cattle feed, cattle can pick up the disease. This is why undercooked beef is the most common method through which people become infected. Only cats that hunt live prey or scavenge cadavers can carry toxoplasmosis. Pregnant women should not be responsible for cleaning cat litter trays, because their developing fetus is at risk from the disease, but if a woman has a positive test to toxoplasmosis before she becomes pregnant, as many do, then there is no risk to the pregnancy.

81. Why do people back away when I breathe on them? Is it because I dribble saliva?

Cat halitosis is quite offensive and is caused by bacteria multiplying in decaying food and infected gums. Cats normally keep their teeth and gums in good condition by chewing on skin and bones. They use their superbly shaped canine or eye teeth to inflict instantaneous death bites on rodents by quickly penetrating between vertebrae in the neck and severing the spinal cord. Their tiny incisors scrape bones, while the larger and sharper molars crush and tear.

Eating canned or soft, moist or even dry cat food doesn't give the teeth and gums the exercise they need. They cease to be self-cleaning and calculus builds up. Once this happens, bacteria multiply, the gums become inflamed, pockets develop between the teeth and food, and debris accumulates. Soon the cat's breath has turned into a potent anaesthetic, knocking out any human who comes in contact with it.

Calculus builds and can become thicker than the tooth it covers. This feels unpleasant and stimulates the saliva glands to work overtime. The result is a smelly, slobbering mouth. All of these problems can be avoided by feeding the cat bones from kittenhood. If that chance has been missed, a cat needs a human to brush her teeth and gums, using a small soft child's toothbrush with something tasty to cats on it: smoked salmon pâté will do nicely. When dribble and odour are present, a cat needs veterinary dental attention before the infection in the mouth gets into the bloodstream.

82. Although my veterinarian vaccinated me against flu, I have been sneezing and my eyes have been runny since I came back from kitty camp. Why?

Cats are routinely vaccinated against cat flu, but they can still come down with signs of flu, even without meeting other cats. If a cat has been to a cattery, he has been in a high risk situation. In addition to the two common viruses in the standard vaccine against flu, there are other causes of flu-like symptoms. One, called Chlamydia, is quite common and can be vaccinated against, but is not always incorporated into standard vaccination procedure. This can be a cause of chronic runny eyes after a cat visits a cattery; it responds to a long course of antibiotics.

A more common cause is also more complicated. One of the causes of cat flu is a herpes virus, a clever germ that manages to hide in the cat's body so totally that no one knows if it is there. If the kitten has already been exposed to the virus before inoculation – and this is common in even the cleanest of catteries – and if the virus is hiding successfully, the vaccine offers no protection.

People think their cat is protected against herpes when in fact he isn't. Not only that, he is actually a silent carrier of the disease, a disease that reactivates under physical or mental stress. Visiting a cattery is a sufficient stress to reactivate the dormant virus. It appears that the cat picked up the disease at the cattery, but in fact he took it there in the first place.

83. All my life I've enjoyed climbing, but now whenever I jump down from the kitchen counter it hurts. Will an aspirin make it better?

Aspirin is an excellent anti-inflammatory but it can be exceptionally dangerous, even lethal to cats. The cause of the pain is the first and most important question to answer.

Cats have such superb bodies that they rarely suffer from the forms of arthritis that afflict humans and dogs. The shoulders are attached to the body only by muscles allowing for lithe dexterity. Sometimes the ball and socket shoulder or elbow joints become inflamed through damage and need time to heal. The ball and socket hip joints can become damaged in a similar way. People first should make sure that a cat is well nourished. A diet of only meat results in a lack of calcium in the bones and an increased risk of damage from something as simple as a jump off the counter. Similarly, a diet of oily fish can result in generalised pain.

If there is a swelling on any bone, a veterinarian should examine it to discover whether there is bone infection from a cat bite or possibly even bone cancer, something that does develop in older cats. Treatment will vary according to the cause of the problem, but a restriction of exercise is almost always one of the most important aspects of nursing.

Aspirin is a marvellous and safe anti-inflammatory in people and dogs. Both of those species have livers that break down aspirin within eight hours, so they can take the drug several times each day. It takes a cat ten times as

96

long to expel aspirin from his body, which is why aspirin should never be given more frequently than every eighty hours, and at a maximum dose of 100 mg per time. More than this can cause serious haemorrhaging.

84. What are my chances of going blind, and if I do is life worth living?

Fortunately the chances of going blind are slim, and for most cats life is still enjoyable. Cats have far fewer inherited causes of blindness than dogs. Cataracts do develop, brought on either through previous eye injuries or more commonly as a result of diabetes. Siamese cats seem most likely to develop cataracts.

Physical injuries to the eyes from cat fights can cause infection and inflammation, which lead to blindness. Sometimes the blindness can be overcome through surgery. Cataracts, for example, can be removed, but only when the rest of the eye is in excellent condition and the surgeon is convinced that vision can be restored.

Coping with blindness can be difficult. Cats must never again go outdoors unless they are accompanied by people, and even then it is very questionable whether it is in their interest to do so. Instead, they should restrict themselves to an indoor life in an environment that never changes. If blindness develops gradually, the sensible cat quickly learns how to navigate around table and chair legs, and can be so successful that humans are sometimes fooled into thinking that she can still see. Her other senses become more acute: she is more aware of vibration and draughts of air. Perhaps most important is that, unlike blind people or even dogs, she can still 'see' through her whiskers. The nerve pathways from the sensitive whiskers to the brain follow the same path as nerves from the eyes. Cats use their whiskers to 'see' at night. Blind cats do the same.

85. *I have gorgeous blue eyes and a beautiful shiny white coat, but I can't hear anything. Is there any connection?*

Yes. There is a genetic relationship between a white coat, blue eyes and deafness. Early breeders of Persian cats selectively bred for the bluest of eyes, but by doing so they encountered a high incidence of deafness. This deafness didn't occur when they bred cats for their natural orange colour.

Blue-eyed, white cats are not always deaf nor are white cats with two different coloured eyes, although it is common for white cats with one blue eye to be deaf on one side. When blue eyes and deafness are linked, kittens are born hearing but lose that ability very quickly. Their brains adapt so magnificently and their other senses become so acute that people sometimes don't realise their cats are deaf at all.

Hearing is an extremely important sense for the hunter, less so for the domestic house cat. A hunter needs good hearing to hear the high-pitched squeaks and rustling sounds of mice. Most domestic cats have their own unpaid slaves that do all their food hunting and preparation for them.

The colour of cats' eyes, incidentally, has nothing to do with the colours they see. Cats see in a variety of shades of green, with a hint of blue. They don't need full colour vision because mice don't change colour as they ripen.

86. Recently, when I counted my toes I was shocked to discover that I have twenty-four of them. Shouldn't there only be eighteen?

Extra toes is a common condition that affects just under ten per cent of all cats, except in the northeastern United States and the Maritime Provinces of Canada, where the incidence is higher than ten per cent.

A typical cat has four toes and a dew claw on each forepaw and four toes but no dew claws on each hind appendage, making a total of eighteen grappling hooks or weapons. Cats with extra digits often have five toes and two dew claws fore and five toes aft, all of them completely functional. Because these redundant toes don't cause any harm, they don't interfere in the life expectancy of cats. Owners of these extra digits also are just as likely to mate successfully as any other cats.

It seems that among the first cats taken to the eastern seaboard of North America by British settlers were a larger than normal number of cats with extra toes. They only had each other to breed with, and because the condition is a simple genetic one, the number of big-pawed cats increased faster within the closed population there than it did back in England where cats had a wider selection of mates. Extra toes should cause no embarrassment and are, in fact, an excellent talking point.

87. I need an operation, but I'm frightened by the anaesthetic. What are the risks of being anaesthetised and what happens when I have surgery?

Anaesthetics have very little risk attached to them; different types are used for various procedures.

Sometimes an anaesthetic-like drug is given simply to keep a cat still while an X-ray or some other minor procedure is carried out. The cat remains awake, but in a devil-may-care state. Otherwise a proper anaesthetic is given by injection into a vein. (Some veterinarians shave the leg so they can see the vein easily.) These drugs have a short duration of action, lasting only minutes, so if a procedure is going to take longer, a mixture of oxygen and inhalation anaesthetic is breathed in through a tube in the windpipe.

Once the cat is asleep, the veterinary assistant shaves hair from the area to be operated on so that it can be thoroughly sterilised. The operation is carried out and stitches are placed either in or just under the skin. The advantage of the latter is that it leaves nothing to be chewed or licked.

Pain killers, antibiotics or other appropriate drugs are then given, and the cat recovers from the anaesthetic in a safe cage. It takes several hours for the anaesthetic to wear off, which is why cats stay at veterinary clinics for several hours before they go home. People should assume that what causes pain in humans or dogs is likely to cause pain in cats too, and should make sure that pain killers are given whenever necessary.

CHAPTER EIGHT

Grooming and Preventive Care

88. I enjoy the comments I receive about my elegant fur. How often should I groom myself and should I ever let others help?

Cats should groom themselves several times daily and, fortunately, every cat carries his own grooming utensil. The backward-facing barbs on the cat's tongue are perfect for combing through fur to remove any dirt or debris.

Because of their elastic dexterity, cats can lick almost all parts of the body. And they are fastidiously tidy. The anal region is usually cleaned after each evacuation, and genitals are cleaned after sex. Using the tongue as a comb, a cat randomly cleans any part of the body that has become dirty. In a much more stylised manner, he cleans his face by applying saliva to a paw, then wiping it like a wet facecloth in ever increasing circles until the entire face and ears are tidied on both sides. The dew claws are used to pull any debris off unlickable body parts.

All healthy short-haired cats are adept at grooming themselves, but long-haired cats with fine hair need help from people. Fine hair matts within hours, so people must groom long-hairs every single day. If a day is missed, matts will form, especially under the legs and on the belly. Once this happens, they can become so entangled that the only way to remove them is by shaving the cat and allowing a new coat to grow.

89. People shampoo themselves when their hair gets dirty. So do dogs. Should I shampoo and if so should I also use conditioner?

A cat's coat is virtually self-cleaning, so as long as a cat routinely grooms herself and as long as her people groom her on the occasions when she is unable to do so, for example when she is ill, shampoos are unlikely to be necessary. There are certain circumstances, when the coat is contaminated with tar for example, or when the skin is excessively crusty and dandruffy, in which a shampoo becomes a necessity.

Dry cat shampoos are certainly a worthwhile option to pursue, if only because cats find these least offensive. Just as a bird takes a dust bath and a dog rolls in clean sand, a dry shampoo can be very effective in cleaning a cat's coat. More difficult to apply are wet shampoos – difficult because many cats try to hold onto the ceiling rather than get wet.

Other than Van cats from Turkey, felines don't naturally take to water. Although cats can swim, they hate getting wet, and hate someone actively wetting them even more. Special cat shampoos are available to treat various coat conditions, and these should be used as infrequently as possible. A conditioner helps to restore natural oils and softness to the coat. People should anticipate that the sight of a water-filled sink will turn a tame tabby into a Bengal tiger.

90. Recently I have become infested with fleas. They don't bother me, but my people are annoyed because the fleas are biting them. How can I get rid of these little suckers?

Any cat that goes outdoors in warm weather is likely to come in contact with fleas. Many species of animals have their own personal variety of fleas. Birds have bird fleas, rabbits have rabbit fleas, dogs have dog fleas, but *everybody* has cat fleas. The nuisance with cat fleas is their proletarian attitude towards life. Although their preferred host is the cat, they are perfectly happy to dine on dogs and humans. In fact, dog and human fleas are quite uncommon. Almost all dog fleas are, technically, cat fleas.

Fleas wait in the grass until a meal walks by, hop on and have a suck. Contentedly, they then nestle into the warmth of the fur and ride back home. Good parasites that they are, fleas realise that a house makes a better home than a lawn. They hop off into the carpet, find a cosy, quiet, warm place to nestle in, meet fleas of the opposite sex and have fun knowing that whenever they want more food their feline restaurant is always at hand.

With their unrestricted taste for mammals, these fleas readily hop onto human ankles too, leaving mosquito bite-like itchy pimples. Because of this lifestyle, in order to eliminate fleas effectively, both the cat and the house must be treated. Special sprays containing a biological substance that prevents flea eggs from hatching should be used on carpets, sofas and anywhere else where fleas might lurk. Proper insecticides – sprays, powders, shampoos, collars, drops or pills – should be used on or given to cats to kill adult fleas.

People should take special care when using insecticides on cats. Many cats find flea collars irritating, so irritating that they lose the hair on their necks. Other cats are inveterate lickers. If flea spray is licked off it can be very toxic, initially causing the cat to dribble saliva but also causing more serious complications, even convulsions. Sprays should be directed against the lie of the coat so that the insecticide gets close to the skin rather than staying on the surface. If concentrated insecticide drops are used, they should be applied directly on the skin, not on the fur, between the shoulder blades so that the insecticide cannot be licked off. Herbal insecticides are certainly worth considering. They are less toxic to the cat. Unfortunately they are less toxic to fleas too.

91. Insecticides are poisonous. Isn't it dangerous for me to be sprayed with chemicals or wear a flea collar permanently and surround myself constantly in an envelope of nerve gas?

It is a question of priorities. Any effective insecticide must be poisonous to fleas if it is to work but cause no side effects to the cat or to humans. When considering the pros and cons of defleaing, the medical problems that fleas bring with them should also be considered.

As has already been mentioned, fleas are the most common cause of skin irritation in cats, and skin irritation is the most common reason that people take cats to veterinarians.

Fleas are the intermediate host for the most common cat tapeworm, and are a potential means by which certain virus infections can spread to cats and people. Therefore it is important to eliminate fleas, not just because they bite cats and people but because they can be the carriers of other diseases. (Remember, it was the common rat flea that carried bubonic plague across Europe.)

Flea collars with easy to break links in case they hook on branches are effective, but care should be taken because some cats are allergic to some collars. Powders are not as effective as other forms of insecticide, and insecticide tablets are a shotgun method of attack that poisons the entire system. The best method of flea control is through spraying, preferably with a pump-action spray, as cats hate the hiss of an aerosol: it sounds like a sabre-toothed tiger to them.

107

92. I noticed when I passed a stool in my litter tray that there were rice grain-sized worms in it. Then, when I groomed myself, I saw one stick its head out of my bottom, then retract. How did I get them, might I have other worms too, and how can I get rid of them?

These rice grain-sized worms aren't actually worms. They are very mobile single segments from a tapeworm, sacs filled with eggs that will be eaten by flea maggots, which turn into fleas, which get eaten by cats, completing the tapeworm life-cycle.

These and roundworms are the most common worms that a cat can carry. Roundworms are often acquired by kittens before they leave the womb. About thirty per cent of kittens are born with them. These worms look like small pearly white earthworms. Their eggs are passed in cat faeces, where, under suitable conditions, they change into infectious larvae that cats swallow, completing their life-cycle.

Tapeworm prevention means flea prevention, as has already been described in question 91. Roundworm prevention initially involves worming the mother cat. She might show no signs of having a worm load, because the roundworm larvae are not in her intestines but are hiding in tiny microscopic cysts in muscles in her body. Pregnancy activates those cysts, and the now mobile larvae travel into the developing kittens, where they take up residence. If a cat has worms, there are highly effective medicines, given either by injection or by mouth, that selectively destroy the worms without causing any

108

side effects such as diarrhoea. Tapeworm medicines dissolve the mouthpiece of the worm. It is then simply digested as food. The most effective roundworm medicines kill all stages in the worm's life-cycle, including the cystic stages.

Unlike dog parasites, there is no evidence to suggest that either cat tapeworms or roundworms are a health hazard to people. The common roundworm is called Toxocara and this is often confused with the microscopic parasite Toxoplasma which *is* a health hazard to people, especially to the fetuses of pregnant women. Toxoplasma is not a worm and is not eliminated by routine worming. Cats can pick up these parasites when they eat prey such as mice and pass them in their faeces. A cat can avoid getting or carrying toxoplasma by eating only wholesome commercially prepared food.

93. No matter how ill I am, I hate taking medicine in any form. I know that sometimes I must, so what is the least objectionable way for me to do so?

The least objectionable way is probably by injection, but that usually involves a trip to the veterinary clinic which can be emotionally devastating. Otherwise medicines come in liquid or tablet form.

Because of their dry, small mouths, and because most tablets are not made specifically for cats, taking pills can be a tricky procedure. It usually requires two people: one to hold the cat and the other to administer the medicine. Sometimes it is best for people to wrap the cat in a large towel or blanket, with only her head sticking out. Her mouth is prised open and turned skywards while the tablet is dropped in. People then shut her mouth and rub under her neck until she swallows. Sometimes a special plastic pill administrator can be used, but this is potentially dangerous, as it can damage the back of a cat's throat.

Liquids are easier to take, although not all medicines come in this form. Using a dropper or syringe, the liquid is simply squirted into the cat's mouth and swallowed. If a tablet does not have a bitter taste, it can be crushed and mixed in a small amount of tasty, strong-smelling food, but generally speaking medicine should not be mixed into a cat's food. She will still taste it and reject it. Some medicines can be mixed in Marmite and smeared on the cat's leg. Natural cleanliness means that the drug will be consumed.

Some pharmaceutical companies have recognised that cats hate medicines. Some drugs, especially antibiotics, now come specially formulated for cats, flavoured with yeast or other substances. Other medicines are paediatric forms of human medicine repackaged and relicensed for feline use. Fortunately, flavours formulated for human infants are accepted by many cats too. Even better, some new medications work on a slow release mechanism. This means that they only have to be taken once a day, in some instances only once every two days.

94. I've heard horror stories that just because we like to scratch a little, millions of cats have all their front toes amputated. Isn't this retribution terribly cruel, and what can I do to prevent it from happening to me?

It is in the nature of cats to scratch, and humans should accept that need if they want to enjoy feline company. It is true that in many parts of the world cats have their claws surgically amputated so that they don't damage the davenport. What happens is that the cat is anaesthetised and the claw is cut off the bone from behind where it grows. The feet are bandaged, and the hapless cat recovers, undoubtedly in pain, and goes home.

Canadian vets have developed what they consider a humane alternative. Instead of removing the claw, the cat is anaesthetised and a tiny ligament below and behind the claw is cut. This is less painful and leaves all the claws intact but they can't grip onto furniture, or trees. An American vet has gone one step further. Observing that women glue artificial nails onto their natural ones, he developed tiny nail shields that can be glued onto cat claws. The claws are cut, a drop of superglue is applied and a nail-shaped nail sheath is slipped on. As the nails continue to grow, these sheaths are replaced every four to six weeks. Wearing these, a cat can still scratch furniture but not damage it. This seems to be a humane way to approach what can be a very expensive problem. As a simpler alternative, cats should have their nails trimmed weekly and be provided with prominent and attractive scratching posts.

112

People tend to forget that cats scratch not to be nasty, but rather to leave marks of their presence. That is why they prefer to scratch furniture in prominent positions, objects like sofas or kitchen tables. The scratched object is obvious to other passing cats, just as the scratched tree trunk on the African savannah is obvious to other passing lions. They also leave paw sweat on the marker as their own personal identity tag.

Cats should expect to be reprimanded when they scratch where they shouldn't. Well-aimed water pistols work nicely. They should be rewarded when they scratch only their own furniture. Liver pâté is an appreciated reward. People should remember that cats don't like their own furniture tucked away in corners: they like their scratching posts in a prominent position where everyone can see them.

95. I love the smells that come from the kitchen when my humans cook, especially when they concoct meaty stir-fries. Will I hurt myself if I hop on the stove and carry out a tasting when they're not looking?

It will be risky. Cats can withstand greater heat than dogs or people, but they still burn themselves. Because they are such inveterate heat seekers, they often singe their whiskers by lying too close to fires. Standing in hot frying pans is just plain dumb.

Many cats do it because, although people find it uncomfortable when their skin touches surfaces over 44°C (112°F), cats don't mind it until the temperature is above 52°C (126°F). They have fewer heat receptors on their paws. That is why they contentedly walk on proverbial or real hot tin roofs.

People should try to dissuade their climbing cats from mounting kitchen work surfaces, and especially from approaching any potentially dangerous area. This usually requires a degree of ingenuity on the part of humans, because more often than not the climbing goes on in their absence. The simplest cat-climbing repellent is to prevent the cat from gaining access to the kitchen when people are out. If this is impractical, then both dangerous and out of bounds surfaces should be cluttered with unbreakable articles like pots and pans in such a way that wherever the cat lands, she is bound to create mayhem. For most, but not all, cats this is sufficient to frighten them from immediately trying to carry out the same escapade again.

96. Each year my people force me into a small box and frighten me out of my wits by taking me to the veterinarian. Is this really necessary?

For both the cat's well-being and his human's peace of mind, yes, it is. A yearly visit to the veterinarian is like a person's five-yearly visit to his or her doctor. A lot can go wrong in the interim.

Yearly visits are good ways of ensuring that cats are routinely vaccinated against the prevalent contagious diseases. Vaccines might include protection against several different causes of cat flu, enteritis, Lyme disease, rabies or leukaemia. It is also a reminder to carry out routine worming.

A yearly examination allows the veterinarian to examine the ears and skin for parasites, check the teeth and gums for infection or damage, and the eyes and nose for discharges. The heart and lungs are listened to, organs in the abdomen are felt, and the anal glands are checked. Just as important, people can ask or answer questions about diet and exercise, age-related changes and behavioural problems. Yearly examinations become more important as a cat matures, simply because the range of potential problems and their severity increases with age. Under some circumstances a house visit might be a preferable way to carry out a cat inspection, but it is almost always better for the examination to be made at the clinic where full facilities are available.

97. It is a costly business getting veterinary attention. Are there any ways that I can get good care but avoid the expense to my humans?

First, prevention is frequently less costly than treatment. Secondly, insurance to cover the cost of treating illness or disease is usually available. Thirdly, when the chips are down and people genuinely can't afford treatment, it is the rare veterinarian who will refuse to treat.

Pet health insurance is available throughout the EC, in Canada and Australia, but only locally in various parts of the United States. In some other countries like Sweden it is so common that many people think it is actually compulsory. The advantage of insurance is that it cancels out what can be the quite significant costs of some treatments. Instead, monthly or yearly premiums are paid, similar to but much smaller than the premiums people pay for human health insurance.

Cats that go outdoors should always be insured. It's a jungle out there. Speeding traffic, slippery roofs and other tough cats make it likely that medical attention will be needed. Housebound cats are protected from these risks and so the likelihood of serious physical injury or of contracting an infectious disease is much less. Even so, housebound cats can suffer from slow-acting virus infections they brought with them and from all the common problems of old age.

At worst, insurance evens out the monthly or yearly expenses for people who live with cats. At best, it means that when there are medical problems people need not worry about the expense, and that the most sophisticated

116

of diagnostic methods and treatments can be used to restore the quality of life of the insured feline.

Cat people fall into several categories. The most attractive group are those who feel responsible for their felines. Regardless of income, they care for their cats, sometimes obsessively so, even disregarding their own well-being. If they can't afford private veterinary fees, in most countries there are animal charities that either directly care for ill animals or will help with the costs.

A less attractive group of people have cats low on their list of priorities. When it comes to the crunch and medical attention is necessary, they would rather spend their money on something else. Because they have sufficient income, they don't qualify for treatment or financial support from charities. Often veterinarians will care for cats that come from these homes but in private will advise the cats to find more attractive humans.

98. I'm a natural animal and I'd prefer to take natural remedies like herbs to get better. Are herbal treatments and homeopathy of any value?

There are no clear answers to this question. Western medicine has evolved from herbal medicine. A very large proportion of the most important drugs that are used today were first discovered in plants. For example, a major drug treatment for cancer comes from blue periwinkles. The major difference between older herbal medicine and its modern relative is that the quantity and purity of the beneficial substances that were very expensive when extracted from plants have become far cheaper when they are chemically recreated.

That does not mean that modern drug therapy is the only way to treat illness. Medicine has become over-dependent on drugs and has used them in a lazy fashion, more frequently and in larger quantities than necessary. The consequence has been an increasing incidence of adverse reactions, and a reaction from some people leading them to look into alternative therapies.

Herbal medicines frequently work, but there can be adverse reactions when large quantities are taken or no response if the wrong herb is used. Homeopathy will certainly cause no harm and may well even be beneficial, just as acupuncture will cause no harm and helps allevi- ate certain forms of pain. In a crisis, however, it seems foolish to disregard standard Western medical methods of prevention or treatment of a disease when these are known to be effective.

99. Completely unexpectedly my feline housemate recently died of a feline AIDS-related disease. He never had sex with a single cat. How did he get feline AIDS, and am I at risk?

Feline AIDS (FAIDS) was only discovered in the 1980s, but it has been around for much longer. It is not spread by sex but through saliva.

Feline AIDS is a virus infection, related to but different from human AIDS and simian AIDS viruses. People cannot catch AIDS from cats nor can cats catch their kind from people. All of these are slow-acting viruses that suppress the body's resistance to disease and to cancer.

The disease is usually transmitted through bites, so outdoor unneutered tom cats are most at risk simply because they frequently indulge in fisticuffs. In some areas up to seventy per cent of feral outdoor male cats test positive, showing they have been exposed to the virus. Exposure does not necessarily mean that a cat will develop a fatal disease, but it certainly increases the chances of later illness.

Just as with human AIDS, there is still no prevention, no vaccine, so cats should practise safe fighting. Avoiding fights, especially with unknown cats, is the best way a cat can protect himself from this serious infection.

119

100. I've heard that cancer is not uncommon in older cats. I've also heard that it is now curable. Is that really true?

The incidence of cancers increases with age, and, yes, some are now treatable and some can be prevented.

Feline leukaemia virus (FeLV) is a slow-acting virus infection which, like the feline AIDS virus, diminishes the body's ability to fight disease. FeLV sometimes allows another virus to cause cancer of the lymph nodes. Cats can be vaccinated against FeLV, which will prevent this type of cancer from developing later in life. Before vaccination, cats should be blood-tested to make sure they are not already carriers of FeLV.

Cancers are still far less common in cats than they are in people or dogs – or even cows. Some are treatable through simple surgery; others respond to drug therapy or sometimes to radiation therapy. These are the same drugs and radiation used on people, but their dosages are considerably lower. Lower dosages are not as successful, but it is unfair that a cat should suffer side effects from treatment if higher dosages are given. The object of either chemotherapy or radiation therapy is to bring the cancer under control without a cat suffering hair loss, nausea or serious constipation. Fortunately, the most common form of internal cancer, lymphoma, responds well to a combination of drugs, and cats often go into long remissions. Technically this isn't a 'cure', because the cancer might return, but it is the closest thing to it.

101. A final question for my humans. Some of them sneeze whenever I'm near. Others itch and have watery eyes, and some even find it difficult to breathe. Can all these problems really be caused by innocent little me?

Yes. Far more people are allergic to cats than to dogs, but there are things a cat can do about it.

People aren't necessarily allergic to cat hair. They are more likely to be allergic to dead skin, dander, that is naturally sloughed. They can also be allergic to a chemical in cat saliva. This is why a cat scratch causes an itchy reaction in allergic people. They are reacting to the saliva on the claws.

Cats should keep their skin in as healthy a condition as possible. This reduces their dander to a minimum. Rather than grooming themselves, they should permit a non-allergic human to rub them down several times daily with a damp sponge. This eliminates the saliva with which they coat themselves. Humans should always brush cats and cut their nails outdoors so that hair, dander and sloughed nails don't settle in the house. Even a cat-free house can make allergic people sneeze if dander is lying around.

When the allergy is really severe and people feel a tightness in the chest, sponging and outdoor grooming isn't enough. In these circumstances, a cat should pack his bags and either move house or go native, moving permanently into the garden. People should install a heater in the garden shed or garage so that the cat has a warm comfortable home. If a cat is willing to camp outdoors, this removes dander from the house, alleviating what,

121

from a human medical viewpoint, can be a potentially serious situation. People sometimes worry that a cat will suffer if he lives outdoors. They forget that this is what cats are so superbly made for. Climbing, lazing in the sun, prowling at night, communing with nature: this is stimulating, exciting and rewarding to cats. If people provide them with medical care and attention, good food, a warm, dry and secure resting place, and occasional tickles, most cats are supremely content to lead a natural feline existence.

List of Questions

123

CHAPTER THREE   **Training**

CHAPTER SEVEN   **Illness and Disease**

128

129

Index